QGIS 3 TUTORIAL FOR BEGINNERS #4

LEARN TO GEOCODE

QGIS Tutorial and
Video Course

Ian Allan

Title Copyright © 2018 by Ian Allan. All Rights Reserved.

All rights reserved. No part of this book may be reproduced in any form or by any electronic or mechanical means including information storage and retrieval systems, without permission in writing from the author. The only exception is by a reviewer, who may quote short excerpts in a review.

Cover designed by

nitty gritty graphics
nittygrittygraphics.com.au

Ian Allan
Visit my website at gis-university.com

First Printing: November 2018
Update: January 2020
Geocode Mapping and Analysis pl

ISBN: 9781657364424

Any description in a table or spreadsheet that can be matched to something on a map is a geocode.

IAN ALLAN

Other AMAZON QGIS Tutorials by Ian Allan 1
About Ian Allan 2
PART 1 – Introduction 5

    What is a Geocode? 5
    Why is Geocoding important? 5
    Geocoding Concepts 7
    How to improve your geocoding hit rate 9
    Mental Maps – Putting Yourself in Someone Else's Boots 10
    Which QGIS Version Should You Install? 13
    How to Make Your QGIS Interface Look Like Mine 14

        Panel Options 15
        Toolbars 16

    Some Things to Install or Acquire 17

        #1: Install the course dataset 18
        #2: Install/update the MMQGIS plugin 19
        #3: Install OpenOffice 20
        #4: Acquire your Google Maps API key 20
        #5 (optional): Install TextPad 21

PART 2: How to format addresses for geocoding 23

    The Generics of Comma Separated Values (CSV) files 26
    Create an address column from street number, name and town columns 27

        Address column creation: Spreadsheet formula example 28
        Address column creation: QGIS SQL Example 30

PART 3: How to geocode your data with the MMQGIS Plugin 35

    How to merge all your geocoding iterations into a single GIS map 36
    Geocoding to Google Maps and Open Street Maps 38

        First understand Google's Terms Of Use 38
        How to format address data for google geocoding 39
        How to Geocode to Google Maps and OpenStreetMap 43

    Geocoding to a Road Centre Line map 51

        Tiger format road centrelines 51
        Formatting your addresses for road centreline geocoding 52
        How to Geocode to a road centreline map 53

    Geocoding to an Address Point map 59

        Formatting your addresses for Address Point geocoding 59

> Cadastral and address point data formats 60
> Spatial considerations for address point maps 61
> Problems with the August 2019 release of MMQGIS 62
> How to Geocode to an Address Point map 64

Conclusion 69
Coupon for the Companion Video Course 71
Glossary of Terms 73

OTHER AMAZON QGIS TUTORIALS BY IAN ALLAN

All my QGIS tutorials are aimed at researchers, college students and professionals who just want to learn the essentials of a mapping tool. Every text is paired with a companion video course hosted on Udemy.com. I teach you, step-by-step, the essential elements of a QGIS task.

GIS 3 for Beginners #1: Getting Started

Learn to use QGIS 3. Navigate the interface. Create a shaded Thematic Map. Learn GIS basics and geospatial analysis

Available on Amazon.com

https://www.amazon.com/GETTING-STARTED-TUTORIAL-COURSE-BEGINNERS-ebook/dp/B07P1FCKZF/

ABOUT IAN ALLAN

I have authored and co-authored fifteen peer reviewed publications. I have worked professionally as a GIS researcher, taught GIS to thousands of students, and worked as a GIS consultant on projects as diverse as the following…

- **United Nations:** Post tsunami strategic planning in Banda Ache
- **Australian Federal Government:** National Broadband strategic assessment.
- **Victoria Australia's Department of Premier and Cabinet:** Housing affordability modelling.
- **Local Government**: Environmental sustainability modelling for planners.
- **Water industry:** Buried water pipe condition modelling and ease-of-digging modelling.

Since the mid 1990's I've been a GIS researcher, teacher and consultant. Over 5000 students have enrolled in my Udemy GIS courses. Here's what some of my students say about my teaching style…

Brian says: *" Ian's course on Geocoding was terrific! He gives you the tools to thoroughly learn how to geocode whether you are an absolute beginner or more advanced in GIS. I have 4 years experience in GIS, but still learnt a lot from his course. I loved the way too how he threw in extra tips like SQL coding, or mind maps to help with geocoding in QGIS, Google Maps or even Open Street Maps. A great course that I intend to refer back to as a keepsake when needed. Thanks Ian!"*

Caesar says: *It was a very good class, just what I needed to get familiar with QGIS.*

Carina says: *"Perfect course to whom have never used…QGIS! Very detailed on the explanations and really generous additional materials to study."*

Umar says: *"Ian is vast and knowledgeable in what he teaches. I would do another course by Ian if he offered it"*

Nathan says: *"Really great introductory-level course! The examples were simple to follow, but also very useful. All the work can be finished while watching the videos, there was no extra work to be done without guidance. I really look forward to the next course by this instructor. - Cheers and well done!"*

Boojhawon says: *"Simple and very clear lectures with the minimum basics/backgrounds to get a taste of what awaits us further and also making us think of what we can do more."*

PART 1 – Introduction

What is a Geocode?

A geocode is a geographical tag that allows a row in a spreadsheet to be mapped. A street address is an obvious geocode. But there are many more geocodes than that.

> *Any description in a table or spreadsheet that can be matched to something on a GIS map is a geocode.*

The following labels are examples of geocodes…

- Soil type on a soil map,
- vegetation community / vegetation type on a vegetation map,
- a building number on a campus map,
- any street address, feature name, grid reference, etc found on a street directory. Don't laugh. Often historical data contains street directory grid references!

Why is Geocoding important?

Many spreadsheets haven't found their way into google maps yet. For you, these spreadsheets might be data you've collected, or historical data that's been sitting on your hard drive for a long time. Old data should not be laughed at. It is often very useful for understanding trends.

Once your data are geocoded, they can be mapped. "Implicit" spatial relationships can then be analyzed as well as the "explicit" queries that you can already make of your table or spreadsheet. Geocoded data can also be used to mend your spreadsheets and improve the quality data in them.

By implicit, for an address you've geocoded, I mean things like...

- How far is it from another address, workplace, hospital, etc.
- Is it within a postcode, suburb, school zone, electoral boundary, etc.
- Is it on soil that's contaminated, on steep slopes, or some other engineering or planning constraint.
- Is it related to some other important map such as a planning area or a hazard map.

Geocoded data can be made compatible with any boundaries you want. Imagine being able to relate your spreadsheets to any census theme at a census map scale of your choosing! Even census boundaries that you've customised yourself. Here's some examples...

- House sales and house rents: Most governments use these data for housing affordability analyses.
    - When house sales and house rents are geocoded, they can be related to census maps. Then median house prices can be compared to median income in a census area.
- Purchasing behaviors: Most supermarkets have loyalty cards
    - When a customer address is geocoded, supermarkets can compare their customer's purchasing behaviors to census characteristics such as household size, income, ethnicity, etc.

The techniques I'm about to show you can also be used for more obscure geocodes. For example a spreadsheet containing details about pollution levels in water bodies...

- When its geocoded to a GIS map of waterbodies, a spreadsheet can be related to other GIS maps such as a GIS map of relevant industrial premises, and then used to assist tracking down polluters.

Geocoding Concepts

Geocoding puts your spreadsheet data on a map and makes it useful. In doing so, your data can be made compatible with datasets it may not have been compatible with before. Suddenly, you can easily relate your spreadsheet to datasets like planning zones, census maps, soil maps, etc.

Geocoding can be easy for some datasets and difficult for others. The MMQGIS geocoding plugin that we'll be using in this course is very forgiving these days. However, it is still important that the addresses in your spreadsheet are formatted consistently, especially for large spreadsheets. That makes errors much easier to track down.

In Table 1, I use 6 variations of the same address to illustrate some common geocoding issues. Notice how subtle the errors can be. To the uninitiated, an error such as an incorrectly formatted ZIP code can take hours to find. In most cases Data Providers do a fairly good job with their GIS address maps, so it will most likely (but not always) be your spreadsheet that has errors to be fixed.

Table 1: An address needs to be formatted with the Street Number and Street Name in the same column. **Notice that only one of the 6 address variations in this table matches the address in the GIS map.** *Even when columns are correctly formatted, subtle differences in address format can prevent an address from geocoding. Other geocodes such as town name and postcode should be in separate columns. Sometimes you can get errors when the columns you're matching to each other are different data types.*

| 1237 Main St Yarmouth 02664 | This is the address attached to the map |

Street Address	Town	ZIP Code	Comments
1237 Main Street	Yarmouth	02664	Won't geocode – "Street" instead of "St"
1237 Main St	Yarmouth	02664	Will geocode – it matches exactly
1237 Main St.	Yarmouth	02664	Won't geocode – full stop after "St"
1237 Main St	Yarm'th	02664	Won't geocode – Town is abbreviated.
1237 Main St	Yarmouth	2664	Won't geocode – Postcode is missing the leading zero.
1237 State Highway 28	Yarmouth	02664	Won't geocode – "State Highway 28" is not on the street map. Local business addresses are often Main Street rather than Route 28 or Highway 28.

How to improve your geocoding hit rate

In this course we're going to use QGIS to turn a street address in a spreadsheet into a point on a map. Ideally the address you're trying to geocode will have a street address in one column, and additional columns such as ZIP code, locality, etc. That way you can improve your geocoding hit rate by **refining** your geographical search. In other words, if an address can't be found in a ZIP code, see if it can be found in a Locality. And if not in a Locality, see if it can be found within some other boundary such as an administrative area, census area, street map grid, etc.

In this lesson we'll be using the MMQGIS plugin. It has three geocoders that allow you to geocode a street address to...

- a Google map or Open Street Map.
- a street line map,
- an address point map,

In places where there are TIGER (US government) format street line maps, or google maps, the MMQGIS geocoders do a really good job.

To be honest, the MMQGIS address map geocoder would be better if it allowed you to *easily* refine your geocoding using different boundaries (ie. a street address in a ZIP code, a Locality, a region, etc). Some commercial GISs have this functionality, and there's at least one GIS that also incorporates semantic and fuzzy algorithms in much the same way that google's geocoder does.

Although MMQGIS's address geocoder is not ideal, when you use GIS frequently, there will always be challenges like this. In this case you just need to be a little inventive with how you tackle the problem. There are two parts to being inventive. You need to...

- Understand your database in terms of mental maps (ie. putting yourself in the boots of the person who entered the data)
- Turn your understanding of your database into a technical solution.

Mental Maps – Putting Yourself in Someone Else's Boots

Mental mapping is the idea that often people write an address that they identify with, rather than the correct address. For example, someone who lives on the border of two suburbs might be more likely to say they live in the suburb they shop in rather than the one they live in. Or a Realtor might advertise a house as being in a near-by suburb that's more attractive to buyers. Or, in the worst case, the person entering the data just doesn't care.

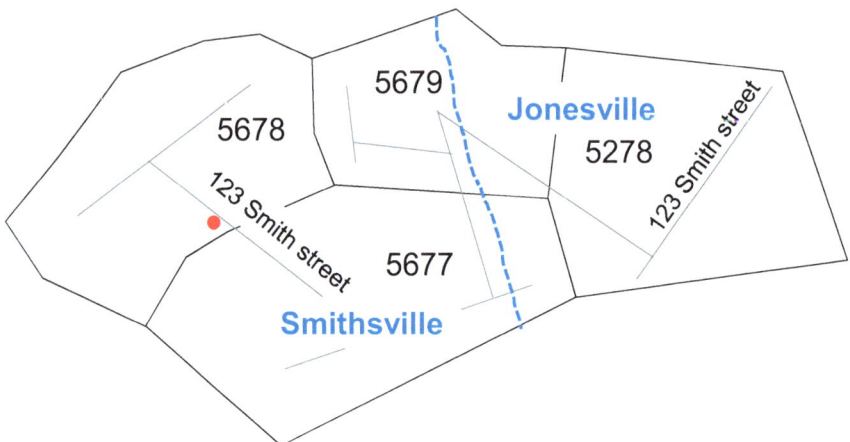

Address	ZIP code	Locality	Description
123 Smith street	5677	Smithsville	Fun resort

Figure 1: This example illustrates two common geocoding problems for "Fun resort" at 123 Smith street Smithsville 5677. Problem 1: The street address (123 Smith street) occurs in two ZIP codes. Problem 2: The ZIPcode in the spreadsheet is wrong (5677 instead of 5678). Solution: geocode using the Address and Locality columns.

An understanding of mental mapping issues can have a noticeable impact on your geocoding success rates, particularly when you're using the street line and address map (non-google) geocoders. Mental mapping is something you must allow for in both the way you format your addresses, and in the way you go about geocoding.

The example in Figure 1 illustrates why it can take multiple attempts to geocode an address. In this case the address of "Fun resort" has been provided by someone who identifies with the *adjoining* ZIP code rather than the resort's *actual* ZIP code. The resort is shown in the table as being at "123 Smith street Smithsville 5677". It could take up to three attempts for this Resort to correctly geocode. Please study this example closely. Understanding the steps to resolve this problem will place you in good stead for dealing with more complicated problems later on in the lesson.

- **Unsuccessful geocoding attempt #1 (duplicate address):** The resort will not geocode by address alone because an address needs to be a unique match. In this case there are two "123 Smith street"s on the map. Solution – introduce a refining boundary such as a ZIP code.

- **Unsuccessful geocoding attempt #2 (wrong ZIP code):** The resort will not geocode using the table's address of "123 Smith street 5677" because its in the 5678 ZIP code on the map. Solution – introduce a different refining boundary – in this case a Locality.

- **Successful geocoding attempt #3 (correct locality):** The resort will geocode using "123 Smith street Smithsville" because the Locality entry in the table matches the Locality of one of the "123 abcd street"s on the map.

The example demonstrates that when you're trying to match an address to a map that has the same street address on different sides of town, the refining boundaries become very important. Even though the ZIP code in the table was incorrect, the Locality was correct and so we could correctly geocode the table to the map. So when the zipcode refining boundaries geocode didn't work, we successfully tried another refining boundary (Locality). In other circumstances we might have to geocode against many additional refining boundaries before a match is found.

Figure 2: Data entry programs can improve the quality of addresses in a system and consequently geocoding hit rates.

If you have a role in your organization's data entry area, you should take note that problems with address matching can be avoided by implementing data input programs that reference a file of actual addresses. This ensures that only valid address can be input into the system (Figure 2).

Which QGIS Version Should You Install?

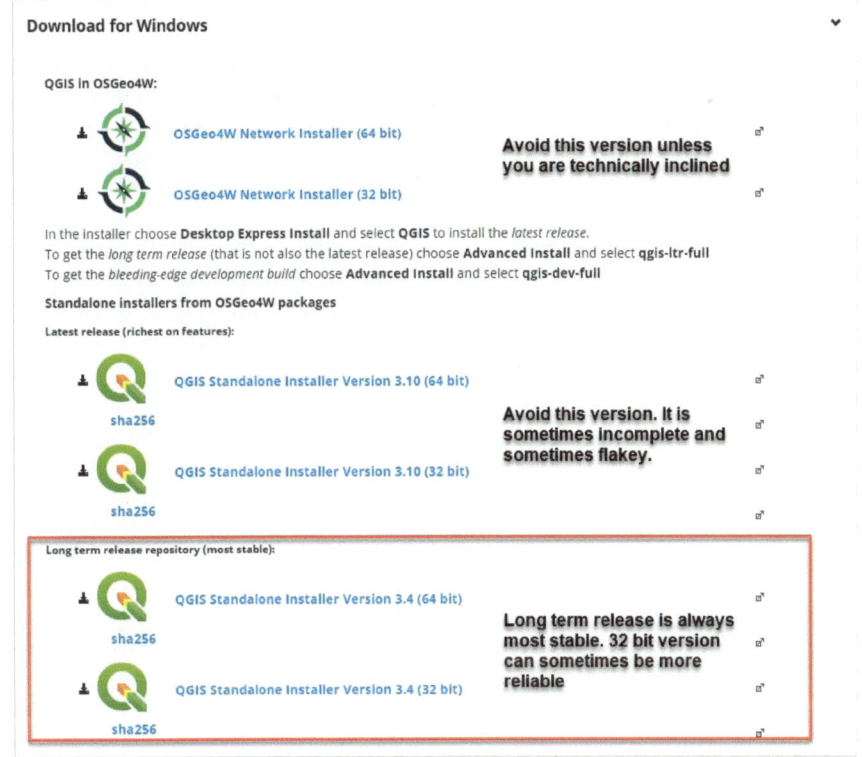

Figure 3: The download page on qgis.org

QGIS is under constant development…

AVOID installing the "Latest release" version. It is the nature of this type of release that it frequently contains errors and omissions. For example, menu items can be missing, or menu items may not work properly.

ALWAYS use the "Long term release" version if you can. It has been checked for errors and omissions, and is deemed suitable for release into corporate environments. This is the version I always try to use. If you have problems with the 64 bit version then try the 32 bit version. The 32 bit version can sometimes be more stable.

How to Make Your QGIS Interface Look Like Mine

You will find that the QGIS interface looks different from installation-to-installation. Sometimes this is because QGIS picks up the way previously QGIS installations were configured, and sometimes, it looks different simply because the default installation changes.

The QGIS interface in my tutorials looks like this. Its best if you get yours looking like mine so you can follow along with the videos more easily. To do that you need to know about QGIS Panels and Toolbars. Both are modified from within the View drop-down menu...

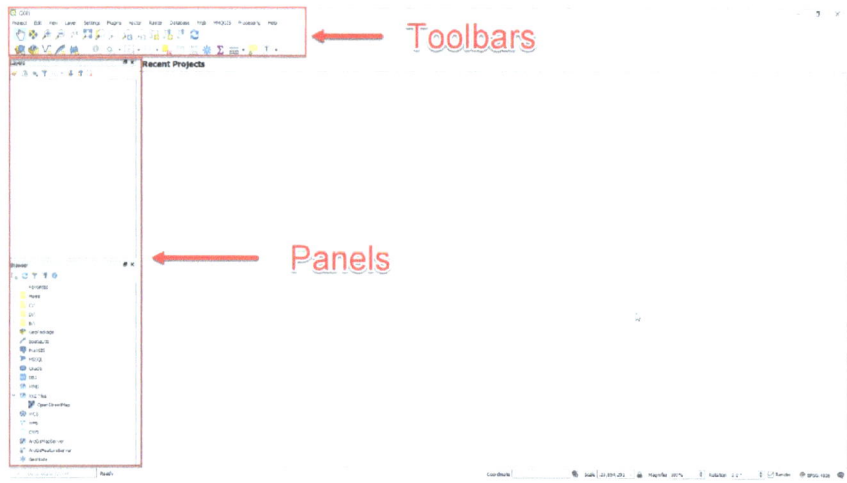

Figure 4: The format for the QGIS interface I'll be using throughout this course

Panel Options

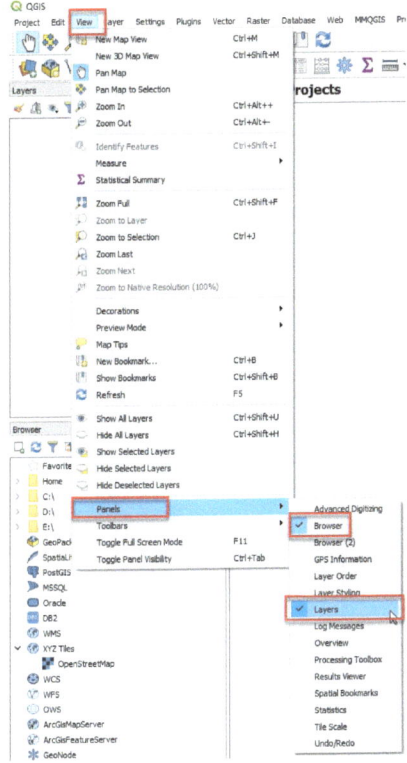

Figure 5: I have both the Browser and Layers panel enabled

I only have the **Browser Panel** and **Layers Panel** active. For me, having other panels open takes up too much screen real estate (Figure 5)

Toolbars

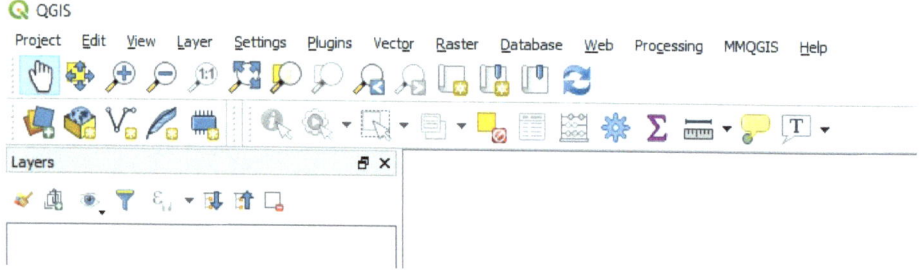

Figure 6: Here's what my toolbar looks like

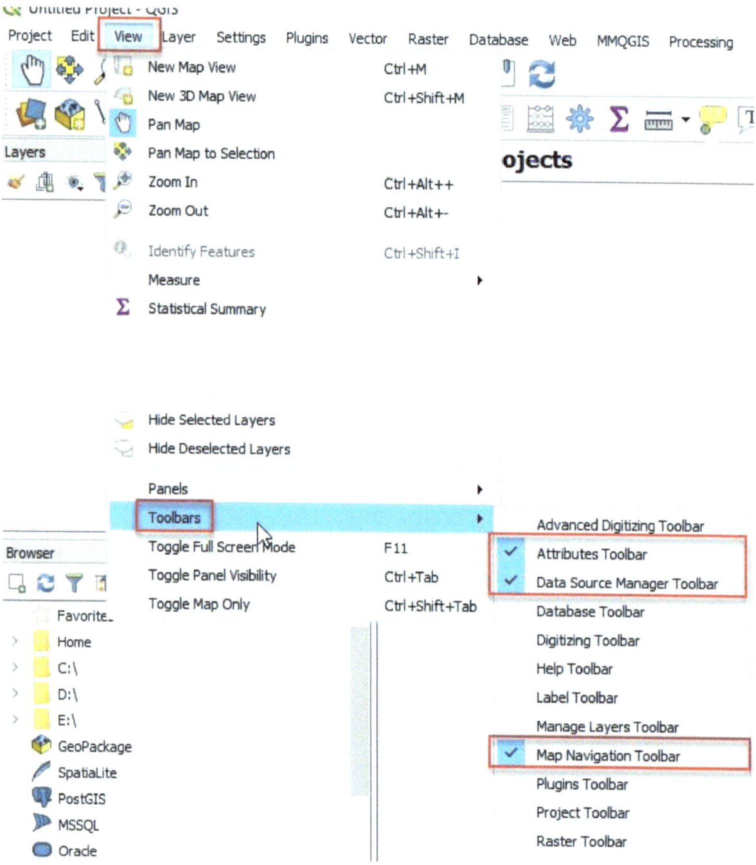

Figure 7: Here's how to configure your toolbar to look like mine

All menu toolbars can be picked up and dragged around the screen. That means that menus can sometimes be in different places than you expect and so they can be difficult to find.

Be sure to grab and drag yours to look exactly like mine. It will make it easier for you to follow along.

Some Things to Install or Acquire

There are five things you need to install or acquire...

Install #1: Install the course dataset. This contains the 44 addresses that we'll be geocoding, the GIS maps we'll be geocoding to, and some sample formulas.

Install #2: Install / update the MMQGIS plugin. This gives us geocoding functionality. In QGIS 3, you should find it already installed in the top menu bar (Figure 8). You'll need to check that this is the latest version. The "how to install the MMQGIS Geocoding plugin" section below tells you how to both install the plugin and check its version if you need to.

Figure 8: MMQGIS in the top menu bar

Install #3: Install OpenOffice. This is an open source equivalent to MS Office.

Acquire #4: Acquire your Google Maps API key. You'll need this key if you want to learn how to geocode to google maps.

Install #5 (optional): Install a text editor to view comma separated value files. You may already have one of these.

#1: Install the course dataset

AMAZON STUDENTS

If you've downloaded this book from Amazon Kindle, you can access the datasets here…

CLICK THIS LINK TO DOWNLOAD THE DATASETS

OR ENTER THE FOLLOWING LINK DIRECTLY…

https://s3-us-west-2.amazonaws.com/datasets-for-amazon-books/Geocoding+Using+QGIS.zip

UDEMY STUDENTS

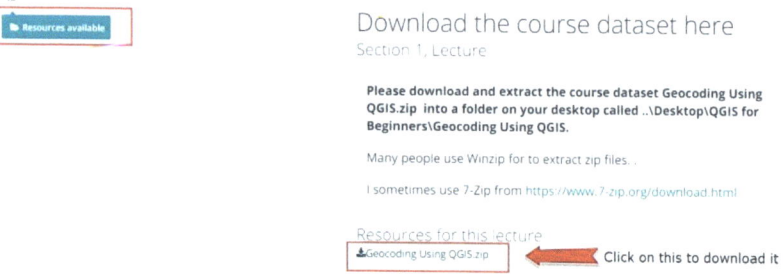

Figure 9: Click on the "Geocoding using QGIS.zip" file to download it. Then extract it using an unzipping program such as winzip or 7-Zip

Please download and extract the course dataset <u>Geocoding Using QGIS.zip</u> into a folder on your desktop called ..\Desktop\QGIS for Beginners\Geocoding Using QGIS. You'll find this in the Udemy lesson called "Download the course dataset here"

Many people use Winzip to extract zip files. I sometimes use the FREE 7-Zip from https://www.7-zip.org/download.html

#2: Install / update the MMQGIS plugin

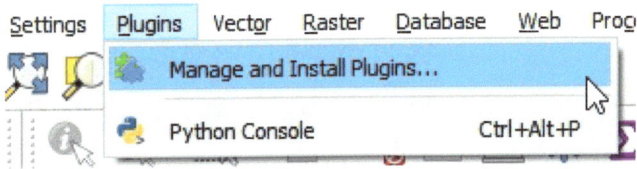

Figure 10: You download Plugins via the Fetch Python Plugins menu.

From the Plugins drop down menu (Figure 10), choose Fetch Python Plugins…

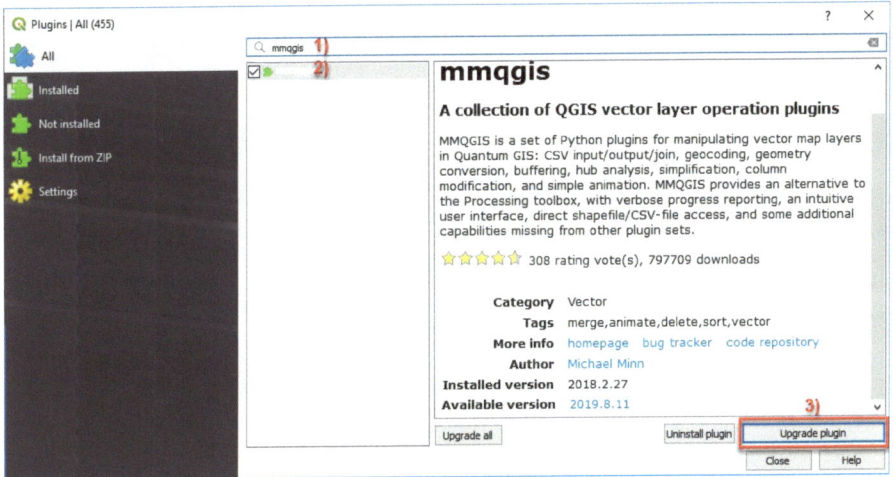

Figure 11: You install / update MMQGIS from the Fetch Python Plugins menu.

Install or upgrade the MMQGIS geocoding plugin as necessary (noting my labels in Figure 11)…

- Type *mmqgis* in the *Filter* box(1).
- Highlight *mmqgis* (2)
- Click the *Install plugin / Upgrade plugin* button (3). When the installation is complete the latest *mmqgis* will be available as a drop-down menu from the top bar. If the plugin is already

installed but there is a newer version available, the box will contain "Upgrade plugin" text.

#3: Install OpenOffice

As much as I like Microsoft Excel, it doesn't seem to work well with the MMQGIS geocoder. Excel's text export format seems to be slightly different to what QGIS is expecting.

There is a free alternative to Excel (and the Microsoft Office suite) called OpenOffice. If you're used to MS Office, then the way OpenOffice works, although not identical, will be familiar to you.

Please download and install OpenOffice from the downloads area of openoffice.org.

#4: Acquire your Google Maps API key

In order to use the Google Maps geocoder in MMQGIS, you'll need a google maps API key. Google's menus to do this change all the time so there's little point for me to do lots of screen captures. Suffice to say…

- Start the process by googling "get google maps api key"
- You will need a gmail address
- When asked to "Enable Google Maps Platform" choose **Places** as the type.
- The 44 addresses in this lesson's database are likely to fall within googles free geocoding plan. As with most google products, this can change. Google require that you enter billing information before they will issue an API key. Then they give you "$200 monthly credit" for free.

- Once you've entered billing information google will take up to around a minute to generate your API key. Copy this and keep it somewhere safe. You'll need to enter it later this lesson.

#5 (optional): Install TextPad

TextPad from textpad.com is my text editor of choice. Although it is free, you have the option of supporting it if you want. For those of you who already have a favourite text editor, Textpad is an optional install.

PART 2: How to format addresses for geocoding

Before you can turn an address in a spreadsheet into a point on a map (geocoding) both...

- the address information in the spreadsheet and
- the address information in the map you're geocoding against

...need to be formatted to be identical to each other.

Each of the three geocoding techniques I'm about to show you require your spreadsheet to be formatted differently. A computer program can't compensate for spelling errors as easily as a person can. For example "str.", "str", "st", "st.", "St", and "street", may seem identical to a human, but they are not identical to a computer.

Even though the MMQGIS geocoding plugin we're using is very good at standardizing and matching address data, there will always be occasions when any geocoding software withing any GIS you use will fail. That's why its important that you have an understanding of how to go about identifying and fixing geocoding errors. Consistently formatted address data...

- are more likely to result in high geocoding rates, and
- they also make it easier to track down geocoding problems.

I originally planned to provide you with a ready-to-go text file with the addresses correctly formatted so that they'll geocode with minimal effort. However I changed my mind. I want you to work hard to geocode the 44 addresses in the teaching dataset. It is good for you to struggle a bit. Unless you do, you won't really understand how truly difficult geocoding can be.

Historical address datasets are the most difficult to gecocode. They were created at a time when people didn't care about address nomenclature issues. They didn't need to because the addresses they were working with were printed onto paper and kept in filing cabinets. They never imagined that their work would be computerized and mapped in a GIS.

For researchers, historical data are very important because they allow change in themes over time to be quantified. As an example, in Australia, housing affordability is a big issue. Some years ago, I was engaged by government to create a GIS map of historic house sales, a contemporary GIS map of all house sales, and assess both against census maps. This allowed Government to better understand the housing affordability crisis.

The addresses in Yarmouth_accom.xls are broken into their constituent columns so you can create your own columns containing addresses custom formatted for whichever of the three geocoding techniques you are using. This spreadsheet is a typical example of an address table that you're likely to come across in any workplace (many will be in much worse shape). If you master the column manipulation techniques I'm about to show you, you should have no problems manipulating your own tables.

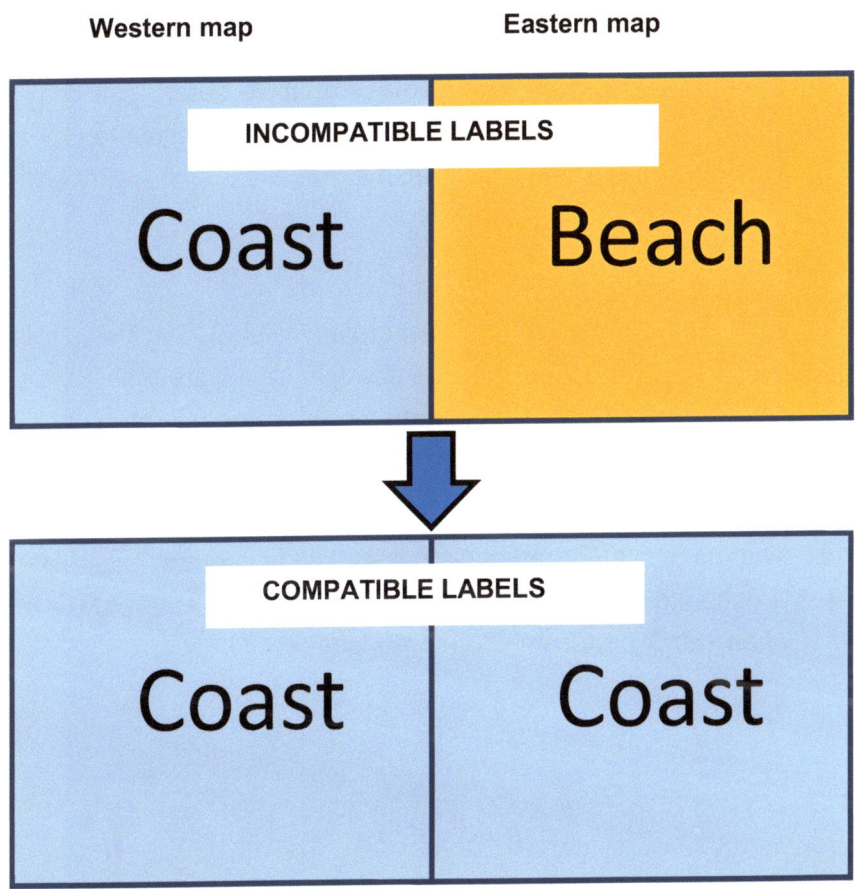

Figure 12: Two adjoining map sheets before the labels have been fixed and after they have been fixed. You fix incompatible text by manipulating text Strings.

To format addresses properly you need to understand string manipulation. Manipulating Strings (text data such as names, addresses and labels) is an extremely important GIS skill to have. Addresses are just one example of Strings you might need to fix. For example, imagine that your study area is within two adjoining mapsheets, but the way thing are labelled in the mapsheets is different. For example, "coast" on one map might be "beach" on another. "Small lake" on one map might be "pond on the other. This is also known as a nomenclature issue.

Adjoining mapsheets need to be made compatible before you can do any GISing. Otherwise your GIS map will have unnecessary legend categories, and queries and calculations you make will be incorrect (see Figure 12).

The Generics of Comma Separated Values (CSV) files

The MMQGIS plugin uses Comma Separated Value (CSV) files. These are text files that you export from a spreadsheet or database. Instead of being organized as neat columns in a spreadsheet, commas separate the data. CSV is a generic database import-export format that is not meant to be read by humans.

All text files used by QGIS must be in a special text format called UTF-8. This is not a big deal. You just need to remember to select UTF- text format when you export from your spreadsheet.

Using Open Office Calc, I want you to…

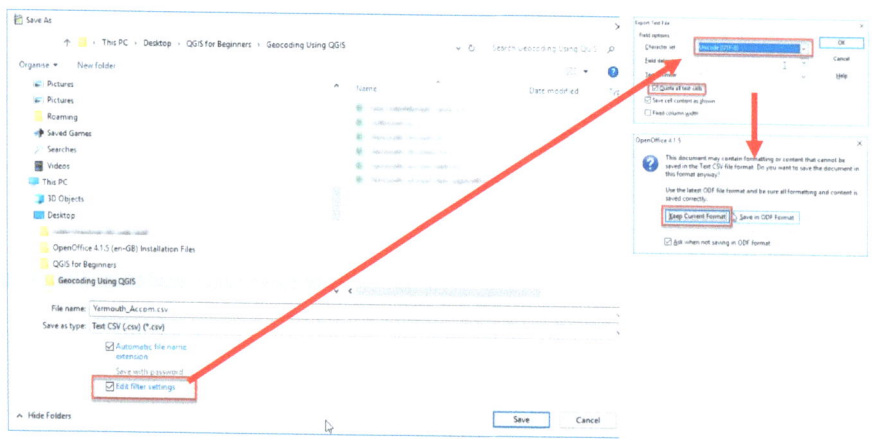

Figure 13: The Save As dialog in OpenOffice. Click Save As from the File drop-down menu and check the "Edit Filter Settings" box. Next check "Keep current format". Finally in the Export Text File box choose Unicode (UTF-8) as the character set and check the Quote all text cells box.

Step 1: Open the Yarmouth_accom.xls spreadsheet then select "Save As" from the file menu and CSV as the format. Save to Yarmouth_accom.csv.

Step 2: Open Yarmouth_accom.csv in a text editor (on MS Windows Notepad or TextPad from textpad.com). Familiarize yourself with how it looks.

```
TEXT:            "Bass River Trailer Park",698,"Willow St.",,"South Yarmouth",2664,"MA","USA"
SMART QUOTES:    "Bass River Trailer Park",698,"Willow St.",,"South Yarmouth",2664,"MA","USA"
```

Figure 14: Text quotes versus SmartQuotes. Smartquotes are the angled quotes that both Microsoft products and OpenOffice use. Although they're pleasing to the eye, they're not "official" text characters and so can cause processing errors. Avoid smartquotes at all costs – use a text editor!

Always avoid looking at your text file in MS Word because its smart-formatting can introduce errors that are hard to track down. If you do use Word, ALWAYS copy and paste your data into a text editor before use it in QGIS. Then in your text editor, use the Find and Replace functionality to overwrite them. The text editor I use is TextPad available for free from www.textpad.com (F8 is launches the Find and Replace functionality).

Create an address column from street number, name and town columns

Address data is often broken into separate columns such as Street Number, Street Name, Suburb, ZIP code, etc. Different geocoding programs require address data to be formatted differently. For example, some require the entire address in one column, and others the street address in one column and refining geocodes such as Zip Codes, Suburb Name, etc, in separate columns.

For this exercise you could choose to manipulate your addresses in one of two ways (or both):

- spreadsheet (OpenOffice Calc), or

- in Quantum GIS.

I would prefer you to master string manipulation in QGIS because what you learn will more easily transfer to other QGIS functionality. However, the QGIS technique is more geeky than the spreadsheet method. I do understand that sometimes you're better off just using the technique you feel most comfortable with and getting things done. When dealing with text files both methods can be tedious. But then again, most facets of GIS database building are tedious!

Address column creation: Spreadsheet formula example

	A	B	C	D	E	F	G
1	Town	State	ZIPcode	STR_num	STR_name	Resulting String	Formula
2	South Yarmouth	MA	2664	698	Willow St	698 Willow St South Yarmouth MA 2664	=D2&" "&E2&" "&A2&" "&B2&" "&C2
3	South Yarmouth	MA	2664	698	Willow St	698 Willow St	=D3&" "&E3
4	South Yarmouth	MA	2664	698	Willow St	698Willow St	=D4&E4
5	South Yarmouth	MA	2664	698	Willow St	698 Willow Street	=D5&" "&E5&"reet"

Figure 15: You can concatenate (join) cells in a spreadsheet to create an address that matches the address format in the map you're trying to relate it to. An = sign signals the beginning of a formula, an & joins two columns and " " delineate text (including whitespaces). Here I show you four examples of ways you might want an address string to look. Note that you have to manually include a whitespace between words. See CodeExamples.xls in the lesson zip file.

Sometimes you can mend minor issues using the Find and Replace functionality in a spreadsheet program. For example you could change all instances of "Str" or "Str." with "Street". Ctrl-f launches the Find and Replace dialog in both MS Excel and OpenOffice Calc. I often use "Find and Replace" functionality for database cleaning in both Excel and Word (its Ctrl-h in newer versions of Word and Ctrl-F in OpenOffice).

In both MS Excel and OpenOffice Calc the String commands are identical. I'm going to demonstrate this in OpenOffice Calc because its UTF-8 characterset CSV export functionality works best.

The idea is to combine the individual columns in your spreadsheet to form an address that's compatible with the address attached to your map. So, if the address attached to your GIS map looks like "698 Willow

St South Yarmouth" then the address in your CSV file needs to be **_exactly_** the same. Even a misplaced dot or whitespace can cause an address to fail geocoding.

The joining (string) command in both Excel and Calc is **&**...

- **Step 1 - Begin your cell with an = sign:** Place your cursor in the top cell of an empty column (row 2 - the one just below your heading row). Begin your cell with an = sign. This tells your spreadsheet program that you're about to write a formula.

- **Step 2 – Insert a cell reference into the formula:** Click on the first cell that makes up your concatenation (eg. house number) and your spreadsheet will insert its reference into the formula.

- **Step 3 – Insert whitespaces:** The resulting string should have a space separating each component so you need to insert that using **&**" "**&**.

Figure 15 contains a number of relevant examples. Note that in row 5, I was able to change **"St"** into **"Street"** by concatenating an **&"reet"** to the end. Look for CodeExamples.xls in the lesson zipfile. It contains the code in the example.

When you've got your formula right then you only need select the cell and hold the left mouse button down while dragging the formula down the page. When you release your mouse button each cell is populated with an address.

Address column creation: QGIS SQL Example[1]

Table 2: You can concatenate (join) columns in a table to create an address that matches the format on your address map. Here I show you four examples of ways you might want an address string to look. The format of the formulas is quite different to those in the spreadsheets. Column names must be in double-quotes, strings (white-spaces, letters, numbers, etc) in single-quotes, and you use double vertical bars to join columns and create spaces to form the completed address. Copy and paste these formulas into field calculator and experiment with them. See CodeExamples.xls in the zip file.

Address components					Resulting String	Formula
Town	State	ZIPcode	STR_num	STR_name		
South Yarmouth	MA	2664	698	Willow St	698 Willow St South Yarmouth MA 2664	"STR_num" || ' ' || "STR_name"|| ' ' ||"State"|| ' ' ||ZIPcode
South Yarmouth	MA	2664	698	Willow St	698 Willow St	"STR_num" || ' ' || "STR_name"
South Yarmouth	MA	2664	698	Willow St	698Willow St	"STR_num" || "STR_name"
South Yarmouth	MA	2664	698	Willow St	698 Willow Street	"STR_num" || ' ' || "STR_name"||'reet'

Another way to build an address is to concatenate it in Quantum GIS. You'll need to hold your breath because it's a bit tricky...

Step 1 – Convert the Yarmouth_Accom csv file to a dbf file: Because you can't edit a CSV file you're going to have to save it into a dbf format file. Do that described to you in **step 1 thru step 3** of the section called "Address formatting for google geocoding: QGIS SQL example" on page 40.

Step 2 – Open the attribute table: Click the Open Attribute Table button to view the table.

---

[1] If ever you need to, you can open dbf and xls (not xlsx) files directly from the Add Vector Layer icon. You can edit a dbf file but you can only view an xls file.

QGIS 3 TUTORIAL FOR BEGINNERS #4: LEARN TO GEOCODE

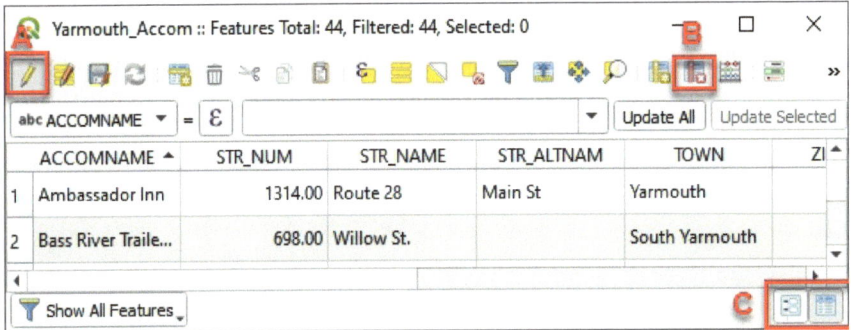

Figure 16: The edit table (A) and field calculator (B) buttons. C toggles between table style view and an individual row view.

Step 3 – Launch the field calculator (Figure 16): Make the table editable (**A**) and then launch the field calculator (**B**). Now we can use the field calculator to add a column made up of any information we want.

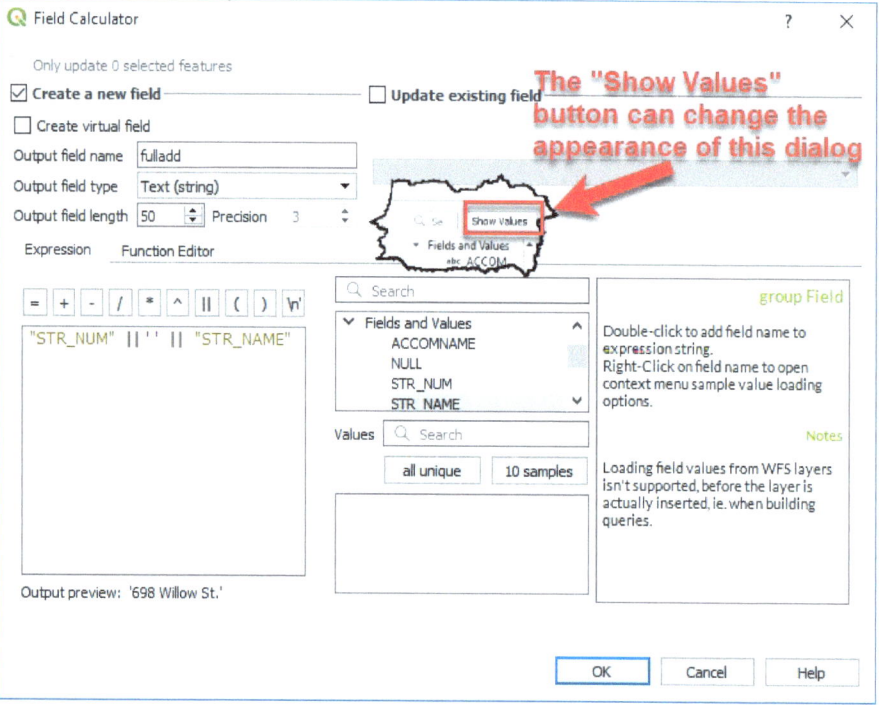

Figure 17: Using the field calculator to create a correctly formatted address. Note that Column Names are enclosed in double-quotation marks ("ColumnName"), text is enclosed in single quotation marks ('white spaces and other text'), and the concatenation (joining) command is double vertical lines (||). NOTE THAT MAX NAME LENGTH IS 10 CHARACTERS

Step 4 – Create an empty column to hold the address: In the *Create new field* area of the dialog box, create a field called "test". Make it *text* with a length of 50.

Step 5 – Insert a cell reference into the formula: In the *Fields and Values* section of the *Function list* box, double-click on the first cell that makes up your concatenation (eg. Street / house number) and QGIS will insert its reference into the *Expression box* at the bottom of the dialog.

Step 6 – Insert whitespaces: The resulting address should have a space separating each component. The joining command is two vertical bars ||. So, to create a space between two address components, type two vertical bars, a single quote, a space, a single quote, and then two vertical bars ||' '|| (ie. "ST_Num" || ' ' || "ST_Name").

Step 7 – Play: In the Field Calculator dialog (Figure 17), play with the commands shown in Table 2. **Table 2** contains a number of relevant examples. Note that in the last row I was able to change **St** into **Street** by concatenating ||"reet" to the end.

We'll create the address columns we'll be using for each of the three geocoding techniques within the relevant sections that follow.

PART 3: How to geocode your data with the MMQGIS Plugin

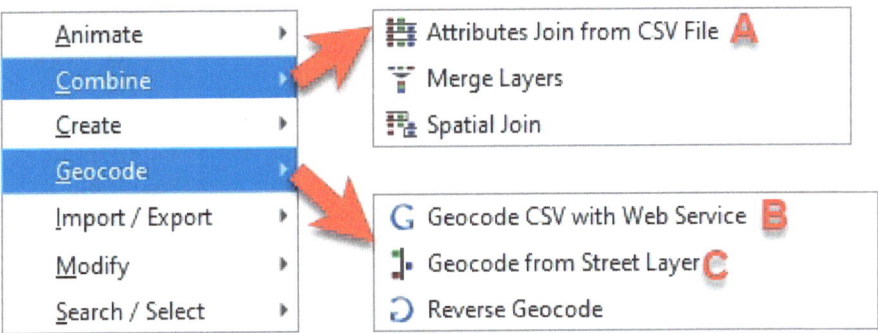

Figure 18: From the mmqgis (ie. Michael Minn QGIS) plugin you can geocode using A) an address map, B) Google maps, C) a TIGER format street line map.

Using the MMQGIS plugin, Geocoding in QGIS is a two-step process. You must...

- **Step 1:** Format the list of addresses in your spreadsheet
    - Format your address labels (street, road, highway, etc) consistently (eg. st, str, street = street)
    - Combine house number and street name into a single column.

To be honest, this step is not as necessary as it used to be. These days the MMQGIS geocoding plugin normalizes address labels in both your text file and your GIS map prior to geocoding, so its hit rate is much better than it used to be. Having said that, it is rare for an address list to geocode successfully on the first attempt. All GIS geocoders can be unforgiving at times, and when tracking down errors, inconsistently formatted addresses are a common cause. That still makes formatting your addresses an important step for training purposes. Each geocoder (MMGGIS has three) requires your address data to be formatted differently.

- **Step 2:** Use one of MMQGIS's three geocoders to map your addresses.

It would be easy for me to just show you QGIS's google geocoder. This is the easiest to use and the most reliable. Google has a very sophisticated algorithm and can match addresses that would not be matched by the other two MMQGIS geocoders. However google's address maps are not available everywhere. In places where google address maps are absent, you need to be able to geocode to GIS maps that are resident on your own computer. By this I mean TIGER format road centreline maps, or address point maps.

How to merge all your geocoding iterations into a single GIS map

Whenever you geocode there will always be addresses that don't geocode for some reason. MMQGIS deals with this issue by placing addresses that don't geocode into a "…notfound.csv" file. Hopefully you can resolve these errors on your next geocoding attempt.

There is a problem that you can end up with a lot address maps that need to be reunited. You do this by using the "merge Shapefiles into One" option. I'm explaining this now so that later in the lesson I can refer you back to here.

You combine multiple geocoding attempts using the "Vector -> data management tools -> Merge Vector layers. Simply select the maps you want to combine and enter a name for the merged map. Then click "run in the background" to merge them.

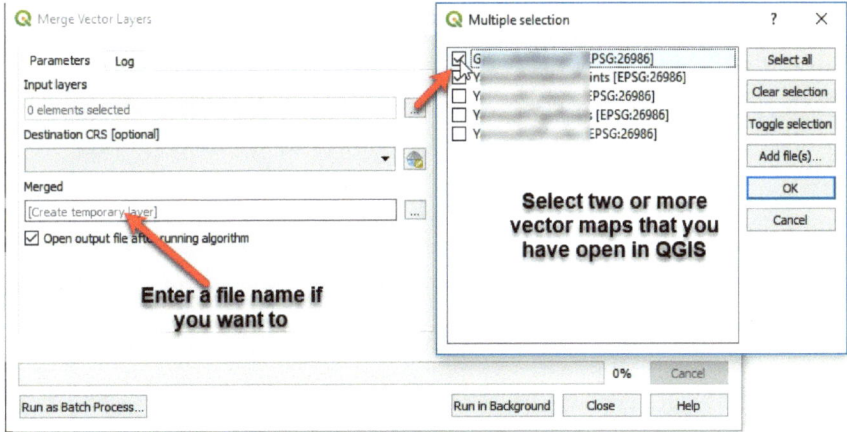

Figure 19: The Merge Vector layer dialog box

Geocoding to Google Maps and Open Street Maps

Google Maps and Open Street Maps are web maps that you can geocode against. Sometimes if you are unsuccessful in geocoding against one of the datasets, its worth trying to geocode against the other.

Google's terms of use are strict while Open Street Maps is Open Data. You should make the effort to read the licensing conditions of each dataset before you use them.

First understand Google's Terms Of Use

When you geocode, you're taking coordinate information from a map and attaching it to a GIS table. When you "take" coordinate information from a table created by someone else, you should be aware that there might be intellectual property and licensing issues with that. This issue is not confined to google. You should always check the way you intend to use your data against your data provider's terms of use.

You need to be careful using the MMQGIS google geocoder because some uses might fall outside Google's Terms of Service.

> *(g) No Use of Content without a Google Map. You must not use or display the Content without a corresponding Google map, unless you are explicitly permitted to do so in the Maps APIs Documentation, or through written permission from Google. In any event, you must not use or display the Content on or in conjunction with a non-Google map. For example, you must not use geocodes obtained through the Service in conjunction with a non-Google map.*

If you are in doubt about whether or not the way you intend to use your geocoded addresses complies with Google's Terms of Service then you should…

- Read the Terms of Use at
  https://developers.google.com/maps/terms#section_10_12

- Get legal advice if you don't understand the Terms of Service.

Face the consequences for ignoring the Terms of Service. These could be, at google's discretion, as minor as suspension or cancellation of your google account, to full-blown legal contest.

How to format address data for google geocoding

ST_Num	ST_Name	ST_AltName	Town	ZIPcode	State	Country	GoogAdd	GoogAltAdd
698	Willow St.		South Yarmouth	2664	MA	USA	698 Willow St.	698
225	Route 28	Main Street	West Yarmouth	2673	MA	USA	225 Route 28	225 Main Street
1261	Route 28	Main Street	South Yarmouth	2664	MA	USA	1261 Route 28	1261 Main Street
167	Old Main St.		South Yarmouth	2664	MA	USA	167 Old Main St.	167
39	Todd Road		Yarmouth	2664	MA	USA	39 Todd Road	39
291	South Shore Dr	South Shore Drive	Y'mouth	2664	MA	USA	291 South Shore Dr	291 South Shore Drive
961	Route 28	Main Street	South Yarmouth	2664	MA	USA	961 Route 28	961 Main Street
8	Harbor Rd, 24 Harbor Rd, 42 Harbor Rd	Harbor Rd	Yarmouth West	2601	MA	USA	8 Harbor Rd, 24 Harbor Rd, 42 Harbor Rd	8 Harbor Rd
512	Route 28	Main Street	West Yarmouth	2673	MA	USA	512 Route 28	512 Main Street
33	Pleasant St		Bass River	2664	MA	USA	33 Pleasant St	33
1237	Route 28	Main Street	South Yarmouth	2664	MA	USA	1237 Route 28	1237 Main Street

Figure 20: For google geocoding, street number (ST_Num) and street name (ST_Name) are combined into one column and other geocodes such as town, state, ZIP code and country are in separate columns.

Addresses should be formatted as number and street name, town, state and country. This gives you the opportunity to explore the idea of mental maps that I discussed earlier. In particular, addresses along the main route into Yarmouth are alternatively known to be on "Main St" and on "Route 28" (see the ST_Name and ST_AltName columns in Figure 20). Sometimes its necessary to use the street's alternate name, and sometimes it isn't. Whether it is or is not is unimportant here because I want to take you through the exercise of creating these two columns.

One way to debug an address that hasn't geocoded is to copy-and-paste it (and variations) into maps.google.com. Google's geocoder is very forgiving and formatting doesn't matter nearly as much as it does for MMQGISs address and street line geocoding routines.

Format address for google geocoding: Spreadsheet formula example

Review the *Address column creation*: Spreadsheet formula example section. Using your preferred spreadsheet program (Microsoft Excel or OpenOffice Calc), complete the following steps... ...

Step 1 – Open the spreadsheet file: Open Yarmouth_Accom.xls in your preferred spreadsheet program

Step 2 – Populate your columns with addresses:

- Create a column called **GoogAdd** (type **GoogAdd** into cell I1)
    - Insert the formula **=B2&" "&C2** into cell I2 and then "drag" the formula down to row 45
- Create a column called **GoogAltAdd** (type **GoogAltAdd** into cell J1)
    - Insert the formula **=B2&" "&D2** into cell J2 and then "drag" the formula down to row 45

Your columns should look like those in Figure 20. If you can't get them to look like that then be sure to watch the video closely.

Step 3 – Save the spreadsheet as a CSV file: Save Yarmouth_Accom.xls and then using the technique I showed you in The Generics of Comma Separated Values (CSV) files(p.26), save it as **Yarmouth_Accom_Google.csv**. This is the file we'll be geocoding to google.

Address formatting for google geocoding: QGIS SQL example

Step 1 – Open the spreadsheet : Open Yarmouth_Accom.xls from the *Add Vector Layer* button that you normally use to open a vector

Shape file. Spreadsheet compatibility is an undocumented feature of the Add Vector Layer dialog box, so to see the xls you will need to select the *All files(*)* option from the list.

Step 2 – Save the spreadsheet file into a format you can edit: In the *Layers* area right-click on Yarmouth_Accom.xls and Save As. From the format drop-down list, choose DBF as the file type. Call it **Yarmouth_Accom.dbf**.

Step 3 – Open the table : Open Yarmouth_Accom.dbf from the Add Vector Layer button that you normally use to open a vector Shape file.

Step 4 – Open the attribute table : Click the Open Attribute Table button to view Yarmouth_Accom.dbf.

Step 5 – Make Yarmouth_Accom.dbf editable :

Step 6 – Launch the field calculator : In here, we're going to create a new column and populate it with address data.

- **Step 6a – Create an empty column to hold the address:** In the *Create a new field* area of the dialog box, create a field called **GoogAdd**. Make it *text (string)* with a length of 50.

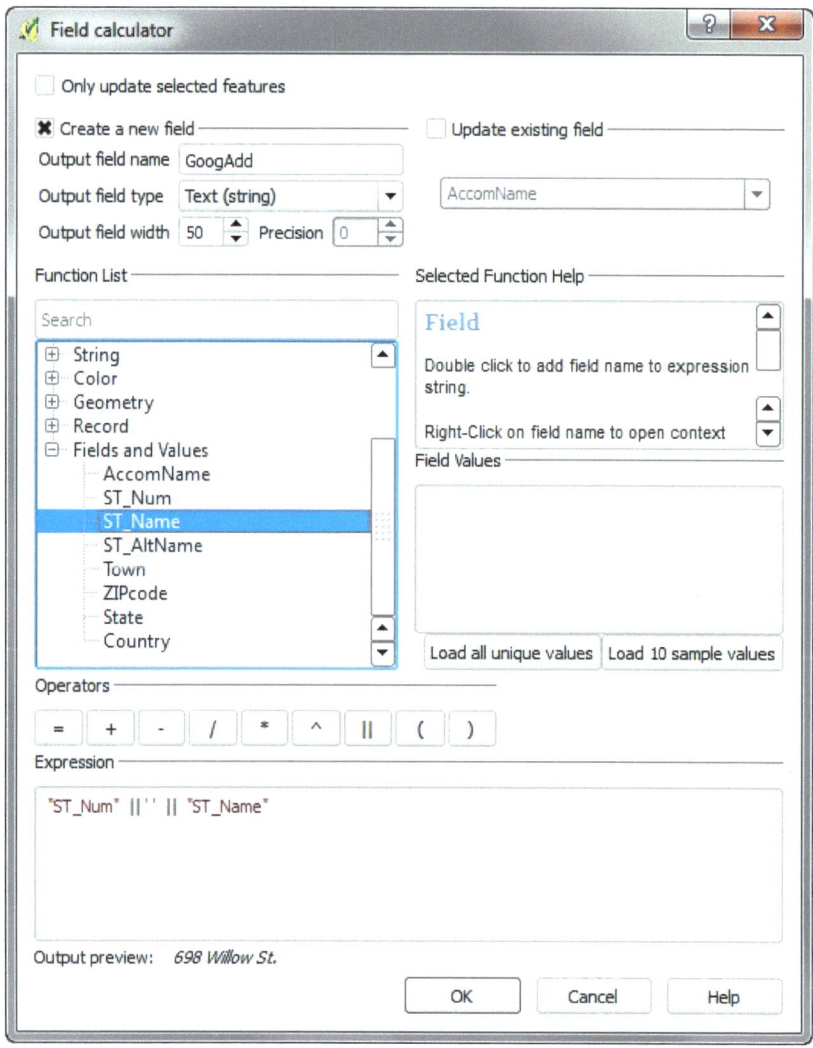

Figure 21: Formula for the Google address

- **Step 6b – Insert the formula:** Use the *Fields and Values* picklist and the *Operators* buttons to insert the following formula into the *Expression* Box ("ST_Num" || ' '|| "ST_Name").

Step 7 – Add the GoogAltAdd column: Repeat step 6, this time for a column called **GoogAltAdd**. In step 6a create a column called **ST_AltName** instead of **ST_Name**.

Step 8 – Save Yarmouth_Accom.dbf :

Step 9 – Convert Yarmouth_Accom.dbf to CSV: Open Yarmouth_Accom.dbf in your spreadsheet program and using the technique I showed you in The Generics of Comma Separated Values (CSV) files (p.26), save Yarmouth_Accom.dbf as **Yarmouth_Accom_Google.csv**. This is the address file we'll be geocoding.

How to Geocode to Google Maps and OpenStreetMap

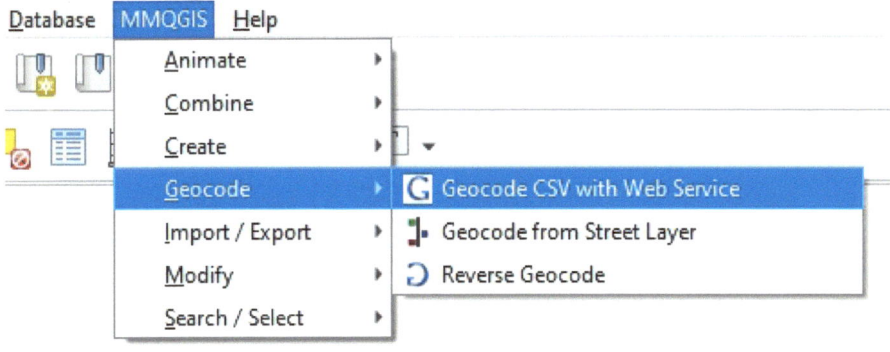

Figure 22: Here's where to find the option to geocode to Google.

You can access the **Geocode CSV with Google / OpenStreetMap** option from the menu path shown in Figure 22. Assuming you have formatted your data as described in the previous section you should take the following steps.

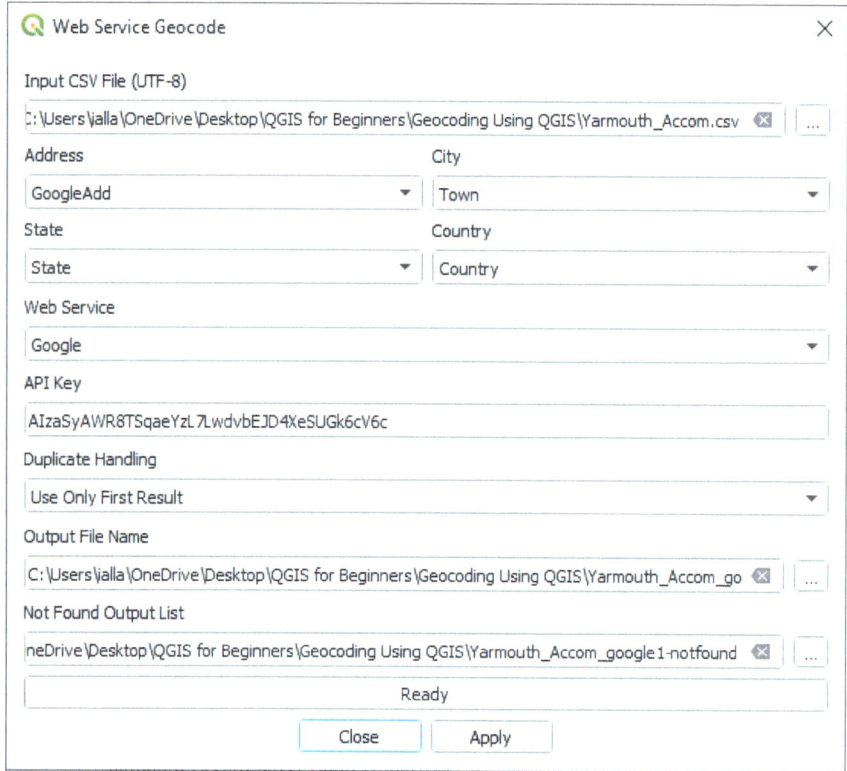

Figure 23: The geocode CSV with Google / OpenStreetMap interface

Step 1 Use address components as shown in Figure 21: Select the fields from the drop-down menu that represent street address, town, state and country (Figure 21). These drop-down menus are important functionality because they allow you to easily substitute alternative address data for those addresses that don't geocode on the first attempt.

Step 2 – fill in the remainder of the form:

- **Output Shape file:** Call this **Yarmouth_Accom_Google1.shp**

- **Not found output list:** Call this **Yarmouth_Accom_Google1-notfound.csv**. This file will contain the addresses that google couldn't match.

Figure 24: The geocode to google progress is in the bottom left corner of QGIS

Step 3 – Click OK: MMQGIS will send your addresses to google's server and then hopefully turn each address into a dot on a map. A progress message displays in the bottom left corner (Figure 24). Yarmouth_Accom_Google1.shp will be added to the QGIS map canvas.

The next part of the process is about fixing geocoding errors. It involves fixing one type of error and then moving onto the next type of error until all the errors are fixed. Some of you might see this as being tedious, and others will see it as a challenging puzzle.

Step 4 – Fix spatial errors: Sometimes addresses will geocode, but to the wrong place. When you inspect Yarmouth_Accom_Google1.shp, do all the dots appear to be logical? Add a map layer such as the cadastre to help you visualize the problem.

- **Click on the erroneous dots** : You might want to use the identify features button to look at the data behind the map object, but realistically the next option is the one you'll end up taking anyhow.

- **Remove the erroneous dots:** This part involves copying the addresses that geocoded to the wrong place and then deleting them from the map. They are shown below as step a thru step d, but there are also a number of sub-steps….

Step 4a – Select the addresses that geocoded to the wrong place:
Select the addresses using the *Select Features by Freehand* tool to select the addresses in error (Figure 25).

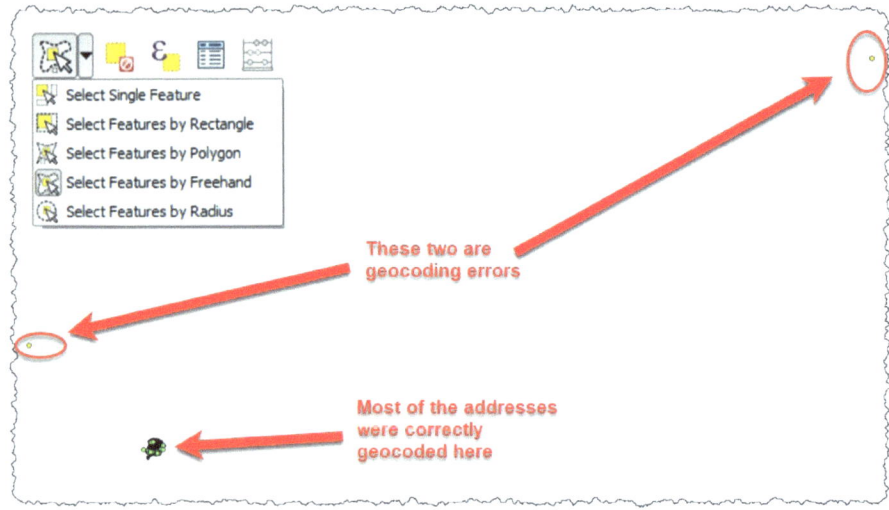

Figure 25: When you've chosen the Select Features by Freehand option you hold the left mouse button down and drag it around the screen. When you release the button, everything in the greyed area is selected.

Step 4b – View the table: click the Open Attribute Table button to view the table behind Yarmouth_Accom_Google1.shp.

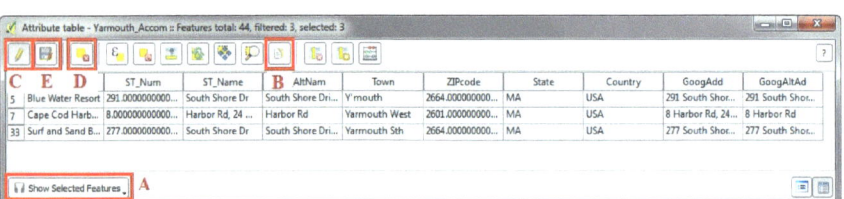

Figure 26: Dealing with addresses that geocoded in error

Step 4c – Delete the addresses that geocoded to the wrong place:
Noting the labels in Figure 26: Dealing with addresses that geocoded in error...

a) Show only the features you selected
b) Copy the rows to the clipboard (for pasting into your spreadsheet program)
c) Make the table editable
d) Delete the three addresses in error
e) Save the changes you made to the table

Step 4d – Paste the rows into a spreadsheet: We're going to fix the addresses if we can. Because maps.google.com contains the addresses that we're geocoding against, if we copy and paste each of the **GoogAdd** values into there, this should give us some clue as to why these addresses did not geocode correctly.

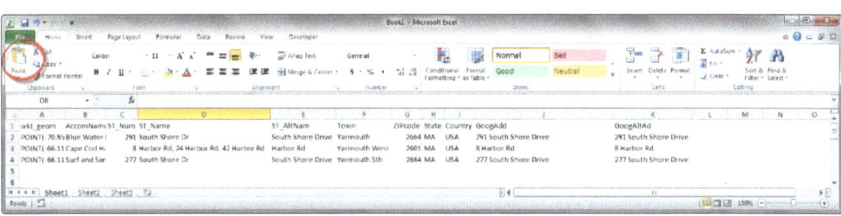

Figure 27: The three addresses that geocoded to the wrong place. Pasting these into maps.google.com will give us some clues as to why they didn't geocode.

- Open a spreadsheet program (Excel or Calc)
- Click the paste button and paste the rows you copied from QGIS into your spreadsheet program.

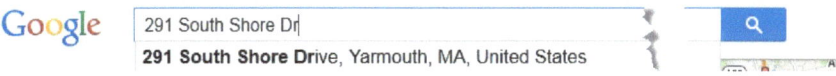

- Copy each GoogAdd value and paste it into maps.google.com. Look for differences between your address and google's version of the same address.

- o **Row 2:** GoogAdd is **Dr** but google's is **Drive**. *Change to Drive.*
- o **Row 2:** Town is **Y'mouth** but google's is **Yarmouth**. *Change Town to Yarmouth.*
- o **Row 3:** GoogAdd has three addresses in it. *Delete the last two addresses.*
- o **Row 3:** GoogAdd is **Rd** but google's is **Road**. *Change to Road.*
- o **Row 3:** Town is **Yarmouth West** but google's is **Yarmouth**. Change Town to **Yarmouth**.
- o **Row 4:** GoogAdd is **Dr** but google's is **Drive**. Change to **Drive**.
- o **Row 4:** Town is **Yarmouth Sth** but google's is **Yarmouth**. Change Town to **Yarmouth**.
- Export the spreadsheet as a CSV called **Yarmouth_Accom_Google2.csv** (see the The Generics of Comma Separated Values (CSV) files section)
- Repeat steps 1 thru 3: Use the file called Yarmouth_Accom_Google2.csv that you just created.

Step 5 – Inspect the addresses that failed to geocode : Click the Add Vector Layer button and open Yarmouth_Accom_Google1-notfound.csv.

- Click the Open Attribute Table button to view the table. Spend some time studying it.

- Inspect the data in the ST_Name column. Fix any obvious problems
- Compare the ST_Name and ST_AltName columns. If there is a pattern of difference between the two columns then this suggests that a second geocoding attempt using the ST_AltName column may solve the problem.

Step 6 – repeat step 1 thru 3: Geocode Yarmouth_Accom_Google1-notfound.csv, this time using the **ST_AltName** column instead.

- Call the outputs **Yarmouth_Accom_Google3-notfound.shp** and **Yarmouth_Accom_Google3-notfound.csv**

Step 7 - repeat steps 4 and 5 for Yarmouth_Accom_Google3-notfound.shp and Yarmouth_Accom_Google3-notfound.csv:

- Copy and paste any failed addresses into maps.google.com. Can you match them in google?

Step 8 - Merge the maps you just created into a single map: Follow the instructions in How to merge all your geocoding iterations into a single GIS map on page 36. Call the merged map **Yarmouth_Accom_Google-WGS84.shp**

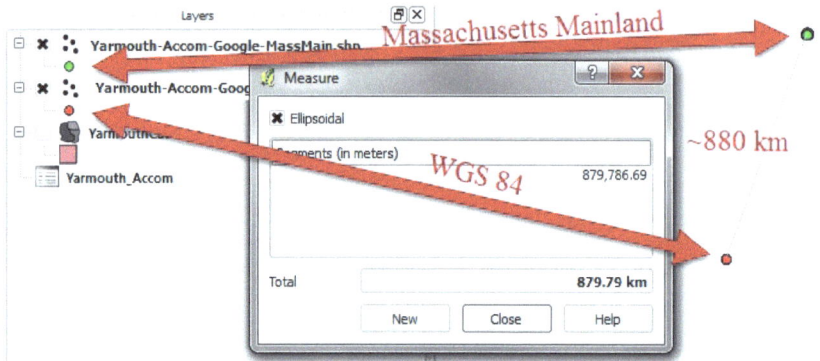

Figure 28: You need to reproject the google geocoded addresses (WGS-84) to the Massachusetts Mainland coordinate system. The difference between these is around 880km

Step 9 – reproject the map: Google returns geocoded items in WGS-84 projection, so if the dots you just geocoded seem to go missing then it's probably a projection issue. You need to reproject your map to be in Massachusetts Mainland projection so it is compatible with your other GIS maps. In the absence of doing this you'll find the "rough" difference between the WGS84 and the Massachusetts Mainland versions of the map is around 880km (Figure 28). To do this you "Save As" a new shape file called Yarmouth_Accom_Google-MassMain.shp in the Massachusetts Mainland coordinate system.

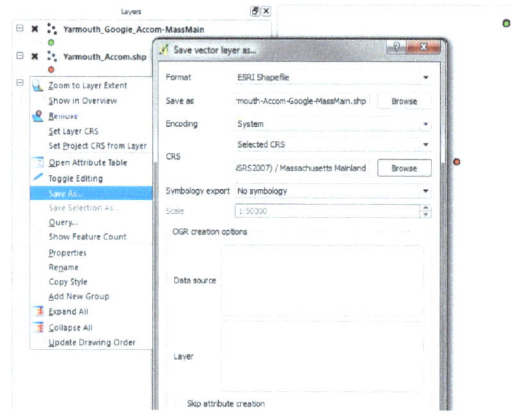

Figure 29: Reproject the google geocodes into the Massachusetts Mainland coordinate system.

Geocoding to a Road Centre Line map

A GIS road centreline map consists of lines on a GIS map representing roads. These are often used as the foundation for commercial street directories. Unlike a dumb street map, a road centreline map consists of many small lines and polylines, each stretching from road-intersection to road-intersection, and each with left-side and right-side address range and ZIP code attached. The MMQGIS geocoding plugin will match the accommodation address in our Yarmouth_Accom.csv file to the address range attached to a road segment in our YarmouthTigerRoads.shp map.

Tiger format road centrelines

	STREET	FROMLEFT	TOLEFT	FROMRIGHT	TORIGHT		Zipl	Zipr	
3517	Lily Pond Dr	32	30	23	13		2664	2664	02(
3518	Great Western Rd	143	135	150	134		2664	2664	02(
3519	Old Main St	182	234	177	231		2664	2664	02(
3520	Woodcrest Ln	1	99	2	98		2673	2673	02(
3521	Norton Rd	70	80	0	0		2673	0	02(
3522	Silverleaf Ln	0	0	0	0		0	0	00
3523	Anastasia Rd	18	10	17	9		2673	2673	02(
3524	Round Dr	25	1	38	2		2601	2601	02(
3525	Mulford St	0	0	0	0		0	0	00
3526	Franklin St	99	9	98	12		2673	2673	02(

Figure 30: TIGER format roads. For the geocoder to work, each road segment must have Street Name, address number range (from left, to left, from right, to right) and ZIP code (left side and right side) information attached to it.

The MMQGIS "Geocode from Street layer" option uses TIGER format road centrelines. If the road centre line map you want to use for your project is not in TIGER format then you'll need to modify its structure to have at least the columns of data shown in Figure 30. Of course, these columns need to be populated with correct data. The geocoder is unlikely to match addresses to any street segment that has data missing.

A detailed discussion about how to format the table behind your road centreline map is beyond the scope of this lesson. It's a complex task because you need to factor in the direction of each street line in order to insert left and right address information. Road centreline maps are usually created by specialist data agencies.

Formatting your addresses for road centreline geocoding

ST_Num	ST_Name	ST_AltName	Town	ZIPcode	State	Country
698	Willow St.		South Yarmouth	2664	MA	USA
225	Route 28	Main Street	West Yarmouth	2673	MA	USA
1261	Route 28	Main Street	South Yarmouth	2664	MA	USA
167	Old Main St.		South Yarmouth	2664	MA	USA
39	Todd Road		Yarmouth	2664	MA	USA
291	South Shore Dr	South Shore Drive	Y'mouth	2664	MA	USA
961	Route 28	Main Street	South Yarmouth	2664	MA	USA
8	Harbor Rd, 24 Harbor Rd, 42 Harbor Rd	Harbor Rd	Yarmouth West	2601	MA	USA
512	Route 28	Main Street	West Yarmouth	2673	MA	USA
33	Pleasant St		Bass River	2664	MA	USA
1237	Route 28	Main Street	South Yarmouth	2664	MA	USA

Figure 31: Yarmouth_Accom.xls - The Street Address geocoder requires the Street Number, Street Name and Zip Code to each be in separate columns. As before, our street name and its alternate name are in separate columns.

Addresses should be formatted as number, street name and ZIP code. This gives you the opportunity to explore the idea of mental maps that I discussed earlier. In particular, addresses along the main route into Yarmouth are alternatively known to be on "Main St" and on "Route 28" (see the ST_Name and ST_AltName columns in Figure 31). Sometimes it's necessary to use the street's alternate name, and sometimes it isn't.

As is the case for the QGIS google geocoder you should debug an address that hasn't geocoded by copying-and-pasting an address (and variations) that hasn't geocoded into maps.google.com.

The MMGGIS geocoder requires our Yarmouth_Accom.csv table to have the columns shown in Figure 31.

Road centreline – spreadsheet formula example

Not required because Yarmouth_Accom.csv is already in the correct format for the MMQGIS Street Layer geocoder.

Road centreline – QGIS SQL example

Not required because Yarmouth_Accom.csv is already in the correct format for the MMQGIS Street Layer geocoder.

How to Geocode to a road centreline map

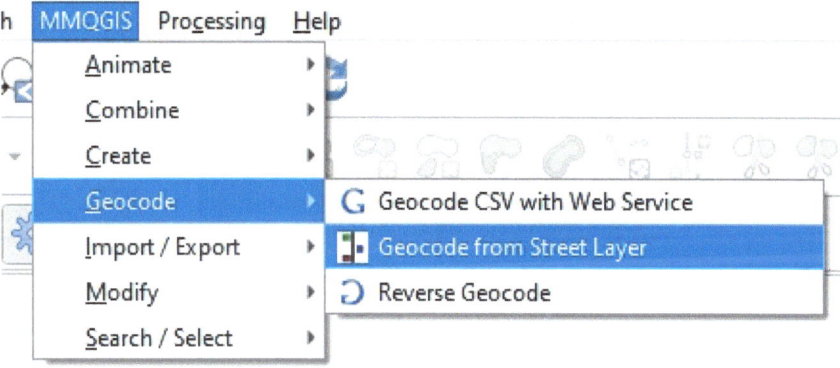

Figure 32: Here's where to find the option to geocode to a Street map.

You can access the **Geocode from Street Layer** option from the menu path shown in Figure 32. The data in our Yarmouth_Accom.csv file is formatted correctly for this, so you're ready to take the following steps. What we're going to do is to tell QGIS to look at each address in the Yarmouth-Accom.csv file and match it to a street in the YarmouthRoads map.

Step 1 – Open the roads map : Open YarmouthTigerRoads.shp from the Add Vector Layer button.

Step 2 – Launch the geocoder: Goto the MMQGIS dropdown menu in the menu bar and launch "Geocode from StreetLayer" (Figure 32).

Step 3 – Fill in the dialog box: In the dialog box (Figure 33), click the Browse button and Open YarmouthAccom.csv.

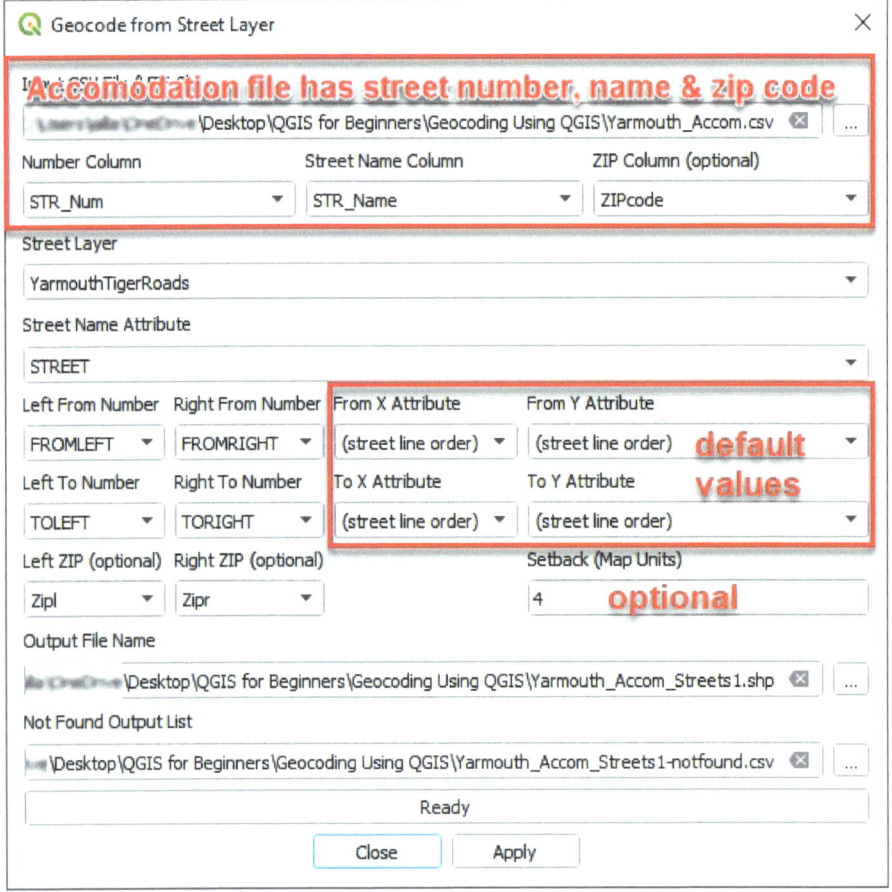

Figure 33: The geocode from Street layer dialog box. Our accommodation file has street number, name and zip code. The roads map has the same data attached to each road segment. The zip code on the left and right sides of the road may be different.

Fill out the dialog box to look the same as that in Figure 33. The following are optional fields…

- **Building setback (optional):** The setback option is important because it ensures that address points are created within property boundaries. Although this approach is questionable for street-line geocoding (street line geocoding is notoriously inaccurate), it does open the door to many powerful spatial data building and query options later.

- **From and To, X and Y attributes (optional):** From X Attribute, To X Attribute, From Y Attribute, To Y Attribute. These refer to the start and end points of street segments. Unless they're attached to your street map table you should ignore these columns.

- Left and Right ZIP codes (optional): Left ZIP, Right ZIP. If you choose (none) in the *Street Layer* area be sure to also choose **(none)** from the ZIP field drop-down list in the *Input CSV file* area.

Select the fields from the drop-down menu that represent street name, street number, and ZIP code. These drop-down menus are important functionality because they allow you to easily substitute alternative address data for those addresses that don't geocode on the first attempt.

- **Output Shape file:** Call this Yarmouth**_Accom_Streets1.shp**
- **Not found output list:** Call this **Yarmouth_Accom_Streets1-notfound.csv.** This file will contain the addresses that couldn't be matched to the street map.

Click OK to geocode the table.

The next part of the process is about fixing geocoding errors. As was the case for the google geocoding exercise, it involves knocking-off one type of error and then moving onto the next type of error until all the errors are fixed. The approach you take to this is dictated more by the types of problems you encounter than by a "recipe" for each of the three geocoding algorithms. Hopefully by the end of this lesson you'll have encountered enough types of problems for you to have a toolkit of approaches to draw on.

Step 4 – Open Yarmouth_Accom_Streets1.shp and fix spatial errors : Sometimes addresses will geocode, but to the wrong place. The YarmouthTigerRoads map should already be on the screen, so when you inspect Yarmouth_Accom_Streets1.shp. It should be obvious at a glance that all the dots appear to be logical. I think this is probably because our streets map only covers our field area.

Step 5 – Inspect the addresses that failed to geocode : Click the Add Vector Layer button and open Yarmouth_Accom_Streets1-notfound.csv.

- Click the Open Attribute Table button to view the table. Spend some time studying it.
- Compare the ST_Name and ST_AltName columns. If there is a pattern of difference between the two columns then this suggests that a second geocoding attempt using the ST_AltName column may solve the problem. That's what we're going to do in the next step.

Step 6 – repeat step 2 and 3: Geocode Yarmouth_Accom_Streets1-notfound.csv, this time using the **ST_AltName** column instead.

  o Call the outputs **Yarmouth_Accom_Streets2.shp** and **Yarmouth_Accom_Streets2-notfound.csv**

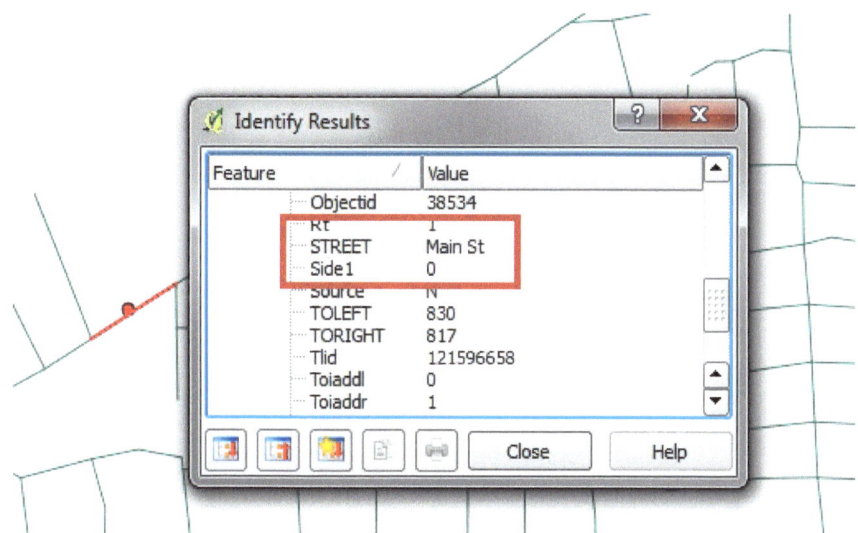

Figure 34: The attributes of a road segment that's next to an address that geocoded to google but did not geocode to the YarmouthRoads.shp map. Compare the STREET name to your addresses to try and understand why the addresses didn't geocode.

Step 7 – Cheat using Google: We could easily take a long time to understand why these addresses didn't geocode. When you're geocoding you often have to be resourceful. One time-saving approach is to compare the addresses we already geocoded using google (Yarmouth_Accom_Google-MassMain.shp) to the addresses we just geocoded to the street map (Yarmouth_Accom_Streets1.shp). In other situations you might choose to use addresses already geocoded using a cadastral or address point map.

Click the Add Vector Layer button and open Yarmouth_Accom_Google-MassMain.shp. Be sure it's the bottom layer so that it's not covering any previously geocoded dots. Click on two or three of the road segments next to address dots that geocoded to google addresses but not to street addresses. Open Yarmouth_Accom_Streets2-notfound.csv in your spreadsheet and compare the format of the address in the STREET column to the way your addresses are formatted in Yarmouth_Accom_Streets2-notfound.csv. Three important differences are shown in Table 3.

Table 3: Three differences between the street address map and the Yarmouth_Accom csv file.

	Yarmouth_Accom_Streets3-notfound.csv	YarmouthTigerRoads.shp
Diff #1	ST_Name is "Route 28"	STREET is "Main St"
Diff #2	ST_Name is "Main Street"	STREET is "Main St"
Diff #3	ST_Name is "South Shore Dr"	STREET is "S Shore Dr"

Step 8 – Fix addresses in your spreadsheet program: Modify the ST_AltName column to contain addresses that are compatible with the addresses in YarmouthTigerRoads.shp and export the file as Yarmouth_Accom_Streets2a-notfound.csv (because Yarmouth_Accom_Streets2-notfound.csv is already open).

Step 9 - repeat steps 2 and 3 for Yarmouth_Accom_Streets2a-notfound.csv: Geocode Yarmouth_Accom_Streets2a-notfound.csv using the **ST_AltName** column.

- Call the outputs Yarmouth**_Accom_Streets3.shp** and **Yarmouth_Accom_Streets3-notfound.csv**

Step 10 - Merge the maps you just created into a single map: Follow the instructions in How to merge all your geocoding iterations into a single GIS map on page 36. Call the merged map **Yarmouth_Accom_Streets.shp**

Step 11: Check for missed addresses: There is a minor bug in the street map geocoder that prevents some addresses from finding their way into the *notfound* csv files. This means that we need to check our source CSV file against our final CSV file. Doing this is a bit tricky because we're using CSV and shape files and so therefore don't have access to QGIS's advanced SQL functionality. That means we need to do our check in Microsoft Excel (Open Office does not have the functionality I'm about to show you).

Geocoding to an Address Point map

A cadastral map is a map of land ownership boundaries. Rather than having address ranges (as with the road centerline map), each land parcel in a cadastral map has an individual address. When you geocode to a cadastral map, your data will be mapped to a cadastral polygon and so is much more accurate than road centreline mapping. Google often uses cadastral and address point maps derived from cadastral maps.

Formatting your addresses for Address Point geocoding

ST_Num	ST_Name	ST_AltName	Town	ZIPcode	State	Country	GoogAdd	GoogAltAdd
698	Willow St.		South Yarmouth	2664	MA	USA	698 Willow St.	698
225	Route 28	Main Street	West Yarmouth	2673	MA	USA	225 Route 28	225 Main Street
1261	Route 28	Main Street	South Yarmouth	2664	MA	USA	1261 Route 28	1261 Main Street
167	Old Main St.		South Yarmouth	2664	MA	USA	167 Old Main St.	167
39	Todd Road		Yarmouth	2664	MA	USA	39 Todd Road	39
291	South Shore Dr	South Shore Drive	Y'mouth	2664	MA	USA	291 South Shore Dr	291 South Shore Drive
961	Route 28	Main Street	South Yarmouth	2664	MA	USA	961 Route 28	961 Main Street
8	Harbor Rd, 24 Harbor Rd, 42 Harbor Rd	Harbor Rd	Yarmouth West	2601	MA	USA	8 Harbor Rd, 24 Harbor Rd, 42 Harbor Rd	8 Harbor Rd
512	Route 28	Main Street	West Yarmouth	2673	MA	USA	512 Route 28	512 Main Street
33	Pleasant St		Bass River	2664	MA	USA	33 Pleasant St	33
1237	Route 28	Main Street	South Yarmouth	2664	MA	USA	1237 Route 28	1237 Main Street

Figure 35: The cadastral address geocoder requires the address data to be in a single column. For the addresses we're geocoding to here, we can re-use the GoogAdd and GoogAltAdd columns we created for the google geocoder.

The MMQGIS "Attributes join from CSV file" geocoder can only match one column of data so at its most basic your HouseNumber and StreetName must be together in one column. This is the column that you geocode. The GoogAdd column in Figure 35 is a good example of such a column.

Boundary geocodes such as post/ZIP codes, town names, state names, etc. should be in separate columns. You may need to use these to create various combinations of "alternative addresses" should the Address column geocoding not produce an acceptable result by itself. To create these alternative address columns, you would use the string functionality you learnt earlier in this lesson. The columns need to be in both your CSV file and your GIS address map. These columns will be along the lines of "GoogAddPlusZIP", "GoogAddPlusSuburb", "GoogAddPlusZIPandSuburb", etc.

A more advanced technique would relate your GIS address map to GIS suburb and GIS ZIPcode maps using spatial query functionality. This would allow you to populate your GIS address map with Suburb and Zip Code columns. You would then use these new columns to create alternative address columns. Exactly how to go about this is beyond the scope of this course but is explained in my Udemy **GIS for Beginners #3: Spatial Analysis using QGIS** (see the bonus area of this course for a discount coupon).

Cadastral and address point data formats

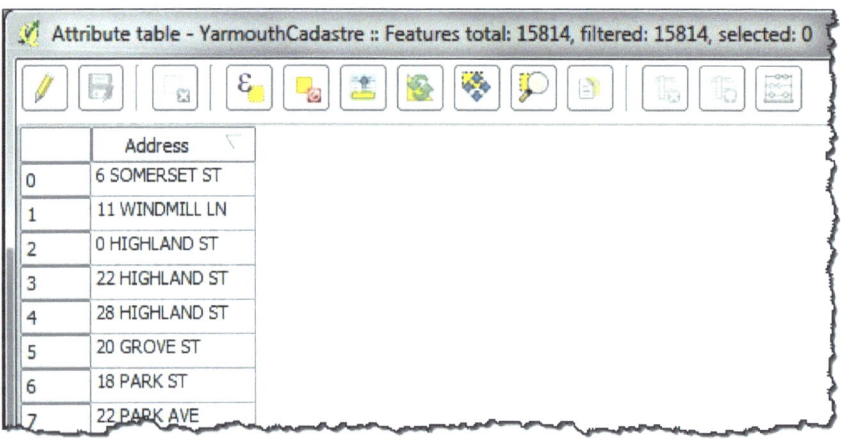

Figure 36: The format of the cadastral map that we'll be geocoding to. The addresses in our Yarmouth_Accom.csv file need to match this identically.

The MMQGIS "Attributes join from CSV file" option matches an address in a single column in a CSV file to an address in a single column in a GIS address map (eg. Figure 36). Unlike the Tiger streetline standard, there is no one way that the address data lying behind each address' geographical object should be formatted. The only guidance I can give you is that your address file must match the format of the GIS address map exactly.

When you do your own projects your GIS address map may well be formatted differently to the one you're about to see. You should have learnt enough in the "How to format addresses for geocoding section" for you to overcome any address formatting issues you might have.

The structure of the table lying behind our GIS map is shown in in Figure 36.

Spatial considerations for address point maps

Some countries have address point databases as well as cadastral databases. These points are positioned either on the centroid (centre of minimum and maximum coordinates) or a para-centroid. In Australia, a para-centroid is a point in the centre of the parcel and set back 4 metres from the road frontage. Para-centroids are a great idea because their positioning is not only predictable, but it also implies which edge of a land parcel fronts the street. In contrast to paracentroids, centroids can be all-over-the-place (Figure 37).

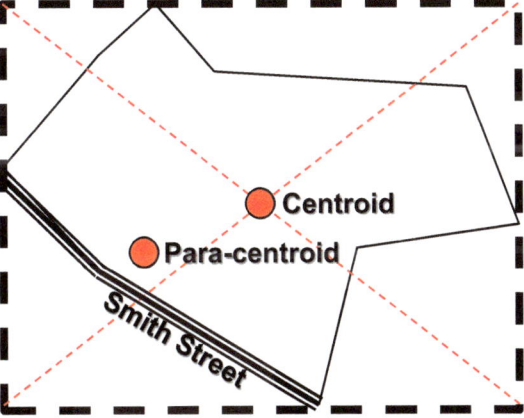

Figure 37: Polygon centroids are found at the intersection of its maximum and minimum coordinates. Sometimes para-centroids are used. In Australia a para-centroid is at the centre of a parcel and 4 metres from the street frontage.

Problems with the August 2019 release of MMQGIS

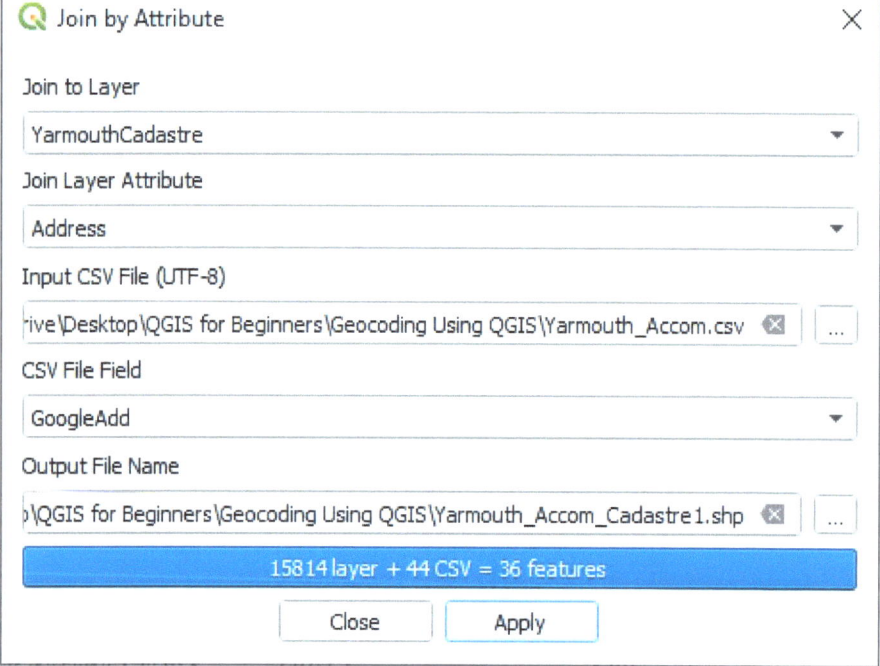

Figure 38: This is the dialog from the August 2019 update of the MMQGIS plugin. The "Not Found CSV Output List" field is absent

There is problem with the August 2019 version of MMQGIS. The "Not Found CSV Output List" field has been omitted. That means that you'll be unable to see which of you addresses did not geocode.

If you would like to do the "full" practical component of this lesson then you will need to install an older version of QGIS and an older version of the MMQGIS plugin. Otherwise, you could just watch the relevant Udemy videos and hope that the next MMQGIS update includes a fix.

First check your version of MMQGIS. It may have been upgraded! If your version has a "Not Found CSV Output List" field then it's OK. If not, here's how you install the older versions of QGIS and MMQGIS...

- Download and install qgis version 3.4.0 Madeira from
  https://qgis.org/downloads/

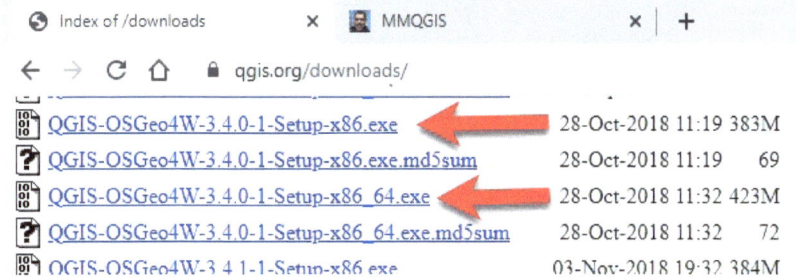

- Download the August 2019 version of MMQGIS from
  http://michaelminn.com/linux/mmqgis/

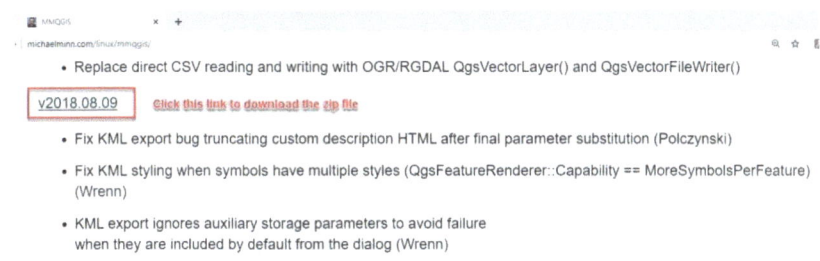

- In QGIS 3.4.0 Madeira, go to the Plugins -> Manage and Install Plugins… menu and install the zip file you just downloaded.

How to Geocode to an Address Point map

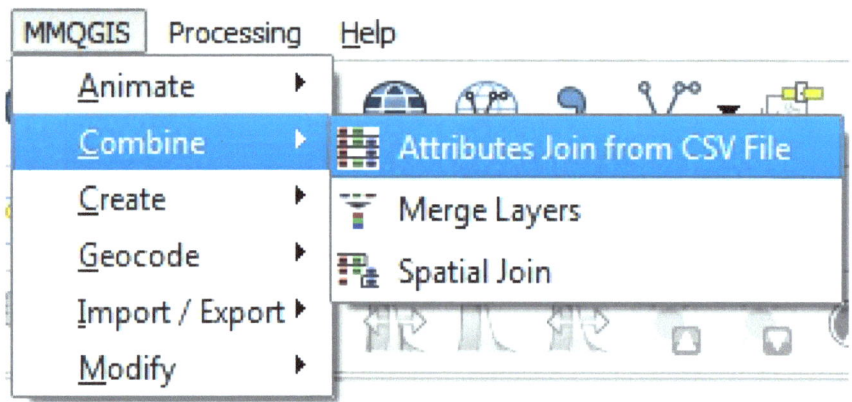

Figure 39: The MMQGIS Address Point geocoding program.

Once you have your desired version of MMQGIS installed, here's how you geocode to an address point.

You can access the **Attributes join from CSV file** option from the menu path shown in Figure 39. The data in our Yarmouth_Accom.csv file is formatted correctly for this, so you're ready to take the following steps. What we're going to do is to tell QGIS to look at each address in the Yarmouth-Accom.csv file and match it to an address in the YarmouthCadastre.shp map.

Step 1 – Open the cadastre map : Open YarmouthCadastre.shp from the Add Vector Layer button.

Step 2 – Launch the geocoder: Goto the MMQGIS dropdown menu no the menu bar and launch "Attributes join from CSV file" (Figure 39)

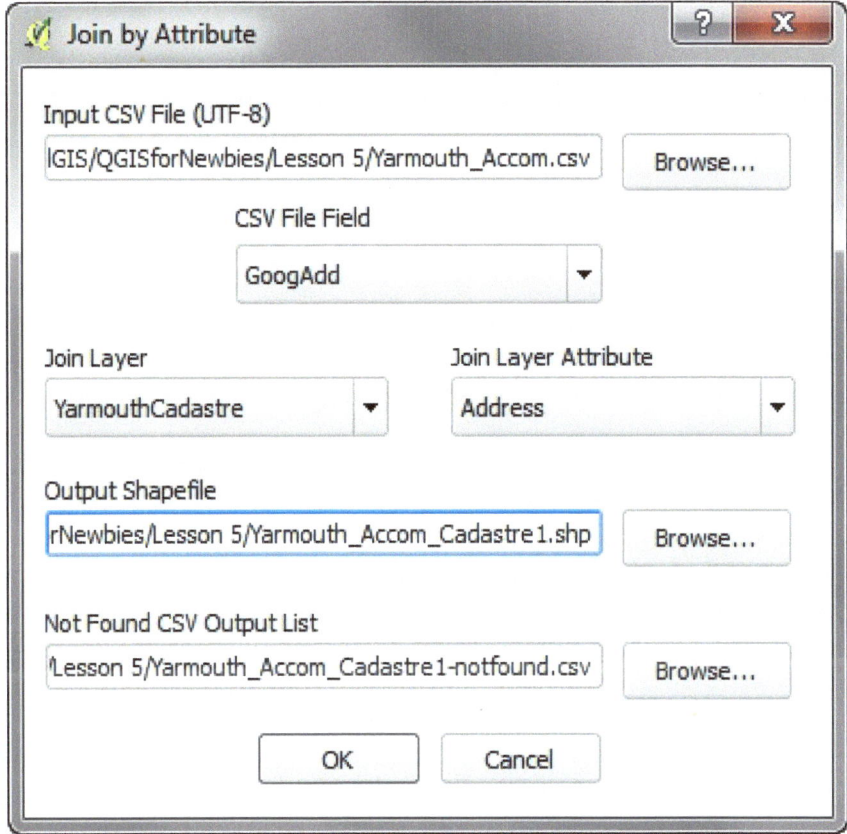

Figure 40: The Join by Attribute dialog box. You use this to geocode to address point information.

Step 3 – Fill in the dialog box: Fill out the dialog box to look the same as that in Figure 40.

- **Input CSV:** Click the browse button and open **YarmouthAccom.csv**. If this geocoder ever crashes, one of the first things to check is that it has been saved using the Unicode (UTF-8) character set. This is an option you can choose in the Save dialog within your spreadsheet. In Open Office you'll find it in the advanced section of the save dialog.

- **CSV File Field:** Choose **GoogAdd**. On the second pass, we'll use the GoogAltAdd column. By this time it will contain addresses that we've repaired.

- **Join Layer:** This is the GIS address map that we want to match our CSV address file to. In this case **YarmouthCadastre** contains our addresses.

Join Layer Attribute: Address is the column in YarmouthCadastre.shp that contains the address data.

Output Shape file: Call this **Yarmouth_Accom_Cadastre1.shp**

Not found output list: Call this **Yarmouth_Accom_Cadastre1-notfound.csv**. This file will contain the addresses that couldn't be matched to the street map.

Click OK to geocode the table.

The next part of the process is about fixing geocoding errors. As was the case for the other two geocoding exercises, it involves knocking-off one type of error and then moving onto the next type of error until all the errors are fixed. The approach you take to this is dictated more by the types of problems you encounter than by a "recipe" for each of the three geocoding algorithms. Hopefully by the end of this lesson you'll have encountered enough types of problems for you to have a toolkit of approaches to draw on whichever approach you choose to take for your own projects.

Step 4 – Open Yarmouth_Accom_Cadastre1.shp and fix spatial errors : As with the other two geocoding techniques, sometimes addresses will geocode, but to the wrong place. When you inspect Yarmouth_Accom_Cadastre1.shp, it should be obvious at a glance that all the geocodes appear to be logical. I think this is probably because our cadastral map only covers our field area.

Step 5 – Inspect the addresses that failed to geocode : Click the Add Vector Layer button and open Yarmouth_Accom_Cadastre1-notfound.csv.

- Click the Open Attribute Table button to view the table. Spend some time studying it.

- Compare the ST_Name and ST_AltName columns. If there is a pattern of difference between the two columns then this suggests that a second geocoding attempt using the ST_AltName column may solve the problem. That's one option. Another option is just to just compare the addresses in Yarmouth_Accom_Cadastre1-notfound.csv to the addresses in YarmouthCadastre.shp.

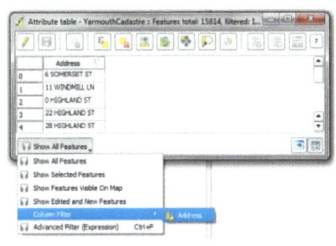 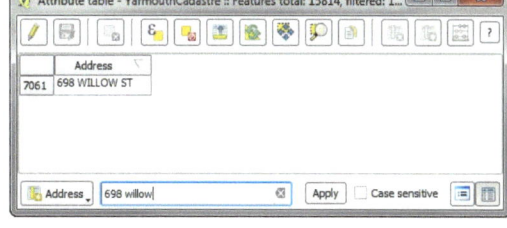

Figure 41: How to filter addresses   *Figure 42: 698 WILLOW ST is in the GIS database. It almost matches our address. If we remove the dot after ST. in our file then our address should geocode. Note that what you type will be case sensitive unless to uncheck the case sensitive box.*

Step 6 – Compare notfound addresses to cadastral addresses:

- Click the Open Attribute Table button to view YarmouthCadastre.shp.

- Open Yarmouth_Accom_Cadastre1-notfound.csv in your spreadsheet
- Type in the failed addresses using the menu path shown in Figure 43 and Figure 41 and when you find the error, type the correct address into the GoogAltAdd column. You'll discover that each of the "notfound" addresses are in the cadastral map, but did not geocode for minor reasons...
    - A dot following the street type.
    - Road types in full rather than being abbreviated.
    - Road types abbreviated rather than being in full.
    - Multiple addresses in the address field

AccomName	ST	Y	GoogAdd	GoogAltAdd
Bass River Trailer Park			698 Willow St.	698 Willow St
Belvedere BandB Inn			167 Old Main St.	167 Old Main St
Blue Rock Golf Resort			39 Todd Road	39 Todd Rd
Cape Cod Harbor Houses			8 Harbor Rd, 24 Harbor Rd, 42 Harbor Rd	8 Harbor Rd
McCormick Cape Cod Cottages			159 Sea View Avenue	159 SeaView Ave
Ocean Mist Beach Resort			97 South Shore Dr	97 South Shore Drive
One Centre Street Inn			1 Center Street	1 Center St
Village Green Motel			37 Seaside Village Road	37 Seaside Village Rd

Figure 43: After inspecting the cadastral addresses, I modified the GoogAltAdd column in Yarmouth_Accom_Cadastre1-notfound.csv to contain addresses that would geocode.

- Export the CSV file with the data enclosed in quotes

Step 6 – repeat step 2 and 3: Geocode Yarmouth_Accom_Cadastre1-notfound.csv, this time using the GoogAltAdd column instead.

Call the outputs Yarmouth_Accom_Cadastre2.shp and Yarmouth_Accom_Cadastre2-notfound.csv

Step 7 - Merge the maps you just created into a single map: Follow the instructions in How to merge all your geocoding iterations into a single GIS map on page 36. Call the merged map Yarmouth_Accom_Cadastre.shp

Conclusion

Well, that's it. Geocoding address data using three techniques!

If your area's GIS infrastructure is good then you are fortunate. You may only need to use one of the three techniques.

Others will be less fortunate. In many places, GIS infrastructure is still being developed. That can mean that different addresses in your table will be found in different GIS databases. If that's the case, then you may need to use all three of the geocoding techniques I've just shown you.

In this text I have focused on address geocoding. The same techniques can be used to geocode to things that are not street addresses.

In the absence of a street address GIS map, there may be more general GIS maps that you can geocode against. For example, geocoding to a map of suburb names, zip codes or some administrative area might be sufficient for the level of spatial analysis you need to undertake.

If you lack GIS maps to geocode against, be sure to treat Google as your friend. Just type a query like "GIS map" followed by the place you're seeking data for. It can be surprising what you find.

Please leave a review so you can help others learn how they can benefit from this book and help me learn how I can better serve my readers.

I hope you found this text useful. Perhaps we'll meet again in one of my other courses?

Thank you and take care!

Ian

Coupon for the Companion Video Course

Enrolment in the Udemy tutorial associated with this book is FREE.

GIS for Beginners #4: Learn Geocoding in QGIS 3: *Learn Geocoding in QGIS 3. Geocode address data from spreadsheets. Geocode to Google, street lines and Address Points.*

Due to recent changes in Udemy policies (free coupons expire after 3 days) you will need to…

- email support@gis-university.com and request a free coupon from me.
- Please attach a copy/screen capture of your receipt to your request.
- Please use the subject heading "AMAZON FREE COUPON PLEASE"

I will return email a coupon. You will have three days to redeem the coupon.

Following are links to all my QGIS tutorials. Feel free to email request a discount coupon for these (support@gis-university.com). The coupon I send you will be for the lowest price Udemy will allow me to at the time (usually somewhere between $10 - $15 USD).

GIS for Beginners #1: QGIS 3 Orientation.

Learn to use QGIS 3. Navigate the interface. Create a shaded Thematic Map. Learn GIS basics and geospatial analysis.

GIS for Beginners #2: Georeference & Digitize in QGIS3:

Learn Georeferencing and Digitizing using Equipment Every Office has in this QGIS tutorial.

GIS for Beginners #3: Spatial Analysis using QGIS:

Learn the Bare Essentials of Spatial Analysis - map overlay, spatial data query and buffering in GIS

GIS for Beginners #4: Learn Geocoding in QGIS 3:

Learn Geocoding in QGIS 3. Geocode address data from spreadsheets. Geocode to Google, street lines and Address Points

Glossary of Terms

Attribute data	Data that relates to a map object. For example, two attributes of a dot on a GIS map might be that… 1. It's a fence post 2. It's made of wood.
Base Map	The map you draw an interpretation on. You create this with • all the features that will assist the person you're working with do their interpretation • all the features that you will need to digitize the interpretation. I recommend that your basemap includes a boundary with known coordinates so you can easily scan, georeferenced and then on-screen digitize from it.
Categorical data	Sometimes called Data Classes. Data that can be expressed as groups. For example Land Use (rural, urban, etc), Vegetation Type (forest, grassland, etc), Habitat Quality (high, medium, low)
Clip board	Standard Windows functionality that allows you to highlight, then copy and paste text and pictures from one computer program (eg. a text editor) to another (eg. QGIS).
Column	The vertical collection of cells in a table. In a table, a column normally has a title which is its reference (for example "DATE"). Data follow underneath the heading (for example, October 17…). *See also, field*
Cross tabulation	The joint distribution of two variables. For example, a cross tabulation of "full time employees" and "teenagers" would reveal all those teenagers who are employed full time.
CSV	Comma Separated Values. A simple form of text file where each column (eg. as exported from Excel) is delineated by a comma. CSV text files are compatible with lots of computer programs.

Database	A collection of tables that are used to describe the project you are working on. In a well designed database, each table will contain information about only one aspect of the project. It is very important that information is stored only once within a database, otherwise you can have problems maintaining data validity and integrity. For example, if a someone's address was stored in two database files and only one file was updated when the person moved house, how would you know which address was correct?.
Decimal number data type:	This data type holds decimal numbers. You can define a field's Length and Precision. The field length is the total length of the field and the field precision is the number of decimal places. For example, a field length of 4 and a field precision of 2 allocates two places for whole numbers and two places for decimal numbers. Valid entries are 99, .99, 99.99. Invalid entries are 999 and .999.
Digitize	*See Digitizing tablet*
Digitizing tablet	A sophisticated electronic tablet on which you attach maps with sticky-tape and trace features with a *puck*. The puck interacts with a dense grid of wires inside the tablet that detect its position and the map features are digitized into the GIS. Later these features get interrogated by sophisticated software that checks its geographical integrity and then turns linework into enclosed polygons. These are not so common these days. Often you can get by by on-screen digitizing. I show you how to do this in my *GIS for Beginners #2: Georeference & Digitize in QGIS* tutorial
DPI	Dots Per Inch. This is scanner speak. The higher the number the more detailed the scan. When I'm georeferencing maps to digitize from, I usually scan them at 300 DPI.
Dynamic map	A map that changes in response to new information. It is always up-to-date. Online weather maps are a good example. Contrasts with a Static map.

Field	A reference to a column of data in a database. Imagine an Excel spreadsheet with a column called "date" and you're some ways to understanding the concept of a Field. However, in a database a field is like a column in an Excel spreadsheet on steroids. Database fields can be referenced by computer programs and GIS queries.
File	*See Table*
Geocode	A text description that can be related to a geographic object (eg. zip codes, addresses, census district names, and local government areas). Imagine you… 1. Have a 300 row spreadsheet. Each row represents a member of a local club. One of the columns contains each member's address. 2. You have a GIS map representing all local addresses You could create a map of your 300 club members by matching the address in each row of your spreadsheet to the GIS map. I show you how to do this in my *GIS for Beginners #4: Learn Geocoding in QGIS 3* tutorial.
Geographical Information System (GIS)	A computer based system for displaying, manipulating and analysing map based information.
GIS	See Geographical Information System.
Ground control point (GCP)	When you're geo-referencing a raster file (ge. a scanned map or an air photo), the scan is in a coordinate system that relates only to itself (ie. row and column numbers), and not to anywhere on the earth (eg. latitude and longitude). In order to relate your scan to an earth position you need to find places on your scan that you can also find in a GIS map. When you digitize these places they become known as *Ground Control Points (GCPs)*. When you've found enough GCPs you run a QGIS program to rectify the image. That turns your scan into an image that can underlay your GIS maps. I show you how to do that in my *GIS for Beginners #2: Georeference & Digitize in QGIS* tutorial.

Term	Definition
Key field	A column in a table that contains a unique identifier that allows a row to be related with a row in another table. Customer Number and Student Number are examples. (see also *Field*). In the example below because the customer numbers in both tables are identical it is possible to print a delivery docket that ensures a Large Refrigerator gets delivered to 10 Smith Street.

Customer Number	Sale Item		Customer Number	Address
1234	Large Refrigerator		1234	10 Smith Street
1235	Small Refrigerator		1235	18 Jones Street
Sales			Customers	

Term	Definition
Line	A GIS object defined by two X and Y coordinates. For example, a section of road.
M value	Refer to horizontal distances (eg buffer widths) in a shape file (see also Z value)
Map layer	A map of a single theme such as cadastre, roads, water features, etc. A GIS maps is created by combining multiple layers. In QGIS, maps can only contain one data type (ie. points, lines, polygons, polylines).
Map object	See object
Mental map	A map that is produced based on someone's understanding of an area. For example, a farmer's mental map of their land will relate to it's productivity and ease of management. An environmentalist's view of the same tract of land might relate to the quality of its wildlife habitat.
Metadata	Information about information. For example, in a census what does employed mean? Working >10, >20, >30 hours each week? Most data custodians have metadata describing their datasets. Often this can be found on the internet or in a library. If all else fails, contact the custodian by email or telephone!
Multi-point	Multiple points that refer to a single row in a table eg a rabbit warren with multiple entrances. Mostly used for sophisticated GIS databases.

Node	*See Vertex*
Nomenclature	The naming convention being used. For example… • having a single spelling for "street" instead of Str, St, Street • having a single way of referring to a coastline. Coast instead of Coast on one map and Beach on another map
Numeric data type	This type of data column only holds whole numbers. Numeric columns are defined to have a length. A length of 1 will hold 0-9, length of 3 will hold 0-99, etc. If you enter a decimal number into a numeric column then it will be rounded to the nearest number. So, 1.2 will become 1 and 1.7 will become 2.
Object	This most often refers to something that is mapped such as a tree, track or paddock. For an explanation of the four GIS object types, see *Points, Lines, Polylines* and *Polygons*.
Point	A location defined by an X and Y coordinate. For example, a power pole or fence post.
Polygon	An area defined by three or more X and Y coordinates. The final coordinate is identical to the first coordinate. For example, a sports oval or a paddock.
Polyline	A location defined by two or more X and Y coordinates. For example, a power line or a fence.
Post Processing	Computer software is used to compare GPS collection to that of a base-station GPS in order to gain centimeter accuracies
Raster map	Also known as a bitmap or a scan. A grid with a different value in each cell. Sometimes cell values can be equated to colours (eg. A photo). Other times, especially in a georeferenced map, the cell values can represent the value of something. For example a height, a slope, or a map category.

Refining boundary.	Identical addresses may exist in multiple places. For example 10 Smith Street exists in the Melbourne suburbs of Carrum, Collingwood and Thornbury. Using "Suburb" as a refining boundary would allow this address to be geocoded to the correct location.
Row	The horizontal collection of cells within a table. That part of a table that contains data. In a GIS, each row usually contains information about a map object.
Segment	This is the line between two vertices/nodes.
Select / Selection	You can select a map object by clicking on it with your mouse either on a map, or within a table. Either method will make a selection in both your GIS map and the attached. When you make a selection, you can choose to do things only to the items you've selected. For example, change its color on the map, or change a value in a column.
Sort	A process by which the values in a data column are ordered either alphabetically or numerically from lowest to highest. In a GIS, this is done using Structured Query Language (SQL).
Spatial	Anything that relates to "space" and can be mapped.
Static map	A map that doesn't change once its been produced. It remains fixed at the point of time in which it was produced. Printed maps are static maps. Contrasts with a Dynamic map.
Structured Query Language (SQL).	The language that a GIS uses to summarize data.
Table	A method of storing data. A table has columns with headings that can be referred to by a GIS, and rows containing data that are "used" by a GIS. A table is normally confined to information about one topic.

Item	Date	Crop
PicnicArea	Oct 17 2010	Sprayed
PicnicArea	Nov 17 2010	Mowed
PicnicArea	Dec 17 2011	Fertilized

Field identifier

} *Column*

Temporal	Anything that relates to time. If we have two maps of the same theme over the same area that have been created at different times then we can map "change". For example, comparing a land use interpretation through time. I show you how to do this in my *GIS for Beginners #1: QGIS 3.4 Orientation* tutorial.
Text	In GIS this means data that are stored in ASCII format. ASCII format data can be read by text editors and can also be read by QGIS. Variations that you're likely to come across are Comma Separated Value text (.csv), straight text (.txt) and Tab Separated text (normally you check a "Tab" box when importing this). Word processor files and database files are NOT text data.
Thematic map	A map about a theme. A theme might be demographic (total population in each postcode), environmental (habitat quality in each postcode), economic (total number of manufacturing plants in each postcode), etc. As you can see, any area can be mapped for numerous themes.
Topological GIS map	A GIS map where objects in the map are intelligent at the database level. For example, the database entry for a polygon will contain information about the polygon adjoining each of its polylines.
UTF-8	A particular form of ASCII (text) that is compatible with many computer programs.
Vertex	When you're on-screen digitizing, each mouse click creates a vertex (sometimes called a node). See also segment.
Windows Clipboard	See *Clipboard*

Z value — Refer to vertical distances (heights) in a shape file (see also M value)